FLORA OF TROPICAL EAS[

LYTHRACEAE

B. Verdcourt

Annual or perennial herbs, subshrubs, shrubs or trees. Leaves simple, opposite or verticillate, rarely alternate; stipules absent or minute. Flowers regular or slightly irregular, hermaphrodite, (3–)4, 6, 8(–16)-merous, solitary to paniculate, sometimes di- or tri-morphic. Sepals united into a tube (hypanthium), lobes valvate, often with small appendages between. Petals free, inserted towards the top of the calyx-tube, alternating with the sepals, folded in bud, or absent. Stamens usually 4 or 8, less often many (10–200) or fewer (1–2), inserted below the petals; filaments sometimes of different lengths in the different forms of flower, usually inflexed in bud; anthers 2-thecous, opening lengthwise. Ovary superior (except *Punica*), sessile or shortly stipitate, completely or incompletely 2–6-locular, rarely unilocular or multilocular (*Punica*); style simple, sometimes of different lengths in different forms of flower; stigma often ± capitate; ovules 2–numerous on axile placentas sometimes not reaching apex of ovary (parietal in one *Ammannia*). Fruit capsular or baccate, opening by a transverse slit or valves or bursting irregularly or ± indehiscent. Seeds numerous, without endosperm.

A family of 26 genera and 580 species. Koehne revised the family in E.P. IV. 216 (1903) but I have not followed his order of genera which appears to me unnatural; *Ammannia* and *Nesaea* are clearly extremely close yet he separates them by 12 genera and places them in different tribes on highly technical characters. S. Graham et al. (J.L.S. Bot. 113: 1–33 (1993)) discuss the classification of the family from a cladistic point of view and confirm the unsatisfactory character of the existing classification. Eight genera occur naturally in the Flora area and a number are also cultivated as ornamentals, one occurring also as an escape. Two species of *Lagerstroemia* have been cultivated.

L. speciosa (L.) Pers. (*L. flos-reginae* Retz.) (queen of flowers, pride of India) (T.T.C.L.: 294 (1949), U.O.P.Z.: 322 (1949), Dale, Introd. Trees Uganda: 47 (1953); Jex-Blake, Gard. E. Afr., ed. 4: 117 (1957)). Tree 7–18 m. tall with much-branched rounded crown; leaves oblong or oblong-lanceolate, up to 30 cm. long, 9 cm. wide; flowers large, mauve or pink, 5–7.5 cm. in diameter, in large elongate terminal panicles; calyx conspicuously ribbed, rusty or grey-velvety; petals 3.2–3.5 × 2.1–2.5 cm., with claw 5–6 mm. long; capsule 1.7–2 cm. long. Specimens have been seen as follows: Kenya, Nairobi Arboretum, 21 Apr. 1953, *Williams Sangai* 524; Tanzania, Lushoto District, Amani Arboretum, 4 Feb. 1971, *Furuya* 229 and Amani, Kiumba, 18 Mar. 1931, *Greenway* 2837; Uzaramo District, Dar es Salaam Botanic Garden, Jan. 1974, *Ruffo* 1493 and 30 Oct. 1979, *Ruffo* 1261; also reported from Uganda, Entebbe, also Zanzibar and Pemba and doubtless extensively grown throughout warmer wetter areas.

L. indica L. (crape myrtle, Indian lilac and sometimes wrongly pride of India) (U.O.P.Z.: 322 (1949), Dale, Introd. Trees Uganda: 47 (1953), Jex-Blake, Gard. E. Afr., ed. 4: 117 (1953)). Shrub 2–3(–6) m. tall; leaves elliptic or obovate, 1–7.5 cm. long, 0.5–3.5(–4.5) cm. wide, obtuse; flowers white, rose-pink, lilac, magenta or mauve in larger terminal panicles; calyx not ribbed; petals 1.6–2.5 × 1.4–1.6 cm., with claw 0.8–1.1 cm. long; capsule 1–1.2 cm. long. Specimens have been seen as follows: Tanzania, Lushoto, 30 Jan. 1971, *Ruffo* 355, 20 Feb. 1971, *Shabani* 667 and 30 Jan. 1981, *Mtui* 43; Morogoro, Nov. 1955, *Semsei* 2393; also grown in Uganda, Kenya and Zanzibar. Masinde reports that the following specimens from Kenya are at EA — Nairobi Arboretum, 20 Mar. 1974, *Mwangangi* 13; Nairobi, Chiromo campus, ICIPE House, 22 Mar. 1983, *Mwangangi* 2446; Nairobi City Park, 24 July 1976, *Kahurananga & Kiilu* 2985; Kwale District, Diani Beach, *Starzenski* in *E.A.H.*10116; he also kindly drew my attention to the existence of a hybrid between these two species of *Lagerstroemia* known as Likoni Hybrid raised by B.L. Perkins in Mombasa, produced by crossing a white-flowered form of *L. indica* (*Perkins* in *E.A.H.* 16283) as female with a blue-flowered form of *L. speciosa* brought from India, Kerala (*Perkins* in *E.A.H.* 16282) as male; material is preserved as *Perkins* in *E.A.H.* 16284.

Several species of *Cuphea* are grown and since one apparently has become naturalised they are dealt with in the main text.

Punica granatum L. (pomegranate) (T.T.C.L.: 464 (1949), U.O.P.Z.: 427 (illustr. on 428) (1949), Dale, Introd. Trees Uganda: 62 (1953), Jex-Blake, Gard. E. Afr., ed. 4: 124, 305 (1957)), a native of Iran and

NW. India frequently put in Punicaceae (but scarcely differing from Lythraceae in its inferior ovary, the calyx-tube adnate to the ovary which is many-loculed with the locules superposed in two series, the lower with axile, the upper with parietal placentation) is widely grown for its edible fruit. Large shrub or small tree, the branches sometimes spiny; leaves opposite, subopposite or fascicled, oblong to lanceolate or obovate, up to 7.5 cm. long, glabrous; flowers orange-red or crimson, showy, ± 2.5 cm. wide; calyx urceolate-funnel-shaped; fruit 3.5–12.5 cm. in diameter, but usually about the size of an orange; seeds numerous, covered with pulp. There are numerous varieties (see Bailey, Stand. Cycl. Hort. 3: 2750 (1939)). Specimens have been seen from Kenya, Mt. Elgon, at 1815 m., Dec. 1930, *Lugard* 300 (apparently away from cultivations); Tanzania, Mwanza District, Igokero, Mbelo, 9 Sept. 1952, *Tanner* 970; Lushoto District, Mangaribi, 21 Apr. 1971, *Mshana* 165 and Kitivo N. For. Res., 27 Mar. 1971, *Shabani* 678; Rufiji District, Mafia I., Chole I., 19 Sept. 1937, *Greenway* 5275. It is widely cultivated in Kenya and Dale reports it as not uncommon in Uganda; it is grown in Zanzibar and Pemba. There are attractive double-flowered forms and a dwarf variety.

Sonneratia sometimes included in the Lythraceae has already been dealt with as a separate family.

1. Trees or shrubs 2
  Herbs or subshrubs, mostly under 1 m. tall (except the
    woody perennial *Cuphea micropetala*) 6
2. Ovary inferior; fruit large, 3.5–12.5 cm. in diameter, the
    seeds covered in juicy pulp; flowers large, bright
    red . . . . **Punica** (see above)
  Ovary superior; fruit smaller 3
3. Leaves black-dotted; flowers somewhat zygomorphic in
    axillary cymes; calyx-tube tubular, longer than broad   1. **Woodfordia**
  Leaves not black-dotted, flowers regular in terminal
    panicles or solitary (rarely paired) in the leaf-axils;
    calyx-tube broader than long 4
4. Stamens numerous; petals 6; capsule not enclosed within
    calyx-tube; cultivated trees and shrubs with showy
    flowers having clawed petals . . . .   **Lagerstroemia**
                                                  (see above)
  Stamens 4, 8, 12 or 18; petals 4 or 6; capsule enclosed or not
    in calyx-tube; indigenous plants with small flowers 5
5. Flowers solitary or rarely paired in leaf-axils; petals 6; fruit
    enclosed by or just exserted from calyx, ± 1-locular,
    circumscissile; silky pubescent littoral shrub of mangrove
    swamp edges, dunes, etc. . . . .   3. **Pemphis**
  Flowers in dense terminal panicles; petals 4; fruit sitting on
    ± peltate calyx, 2–4-locular, indehiscent or bursting
    irregularly; glabrous shrub not confined to littoral
    areas . . . .   4. **Lawsonia**
6. Calyx-tube usually ± curved, over 1 cm. long, gibbous at the
    base, oblique at the throat; cultivated and one
    naturalised . . . .   2. **Cuphea**
  Calyx-tube symmetrical, up to 5 mm. long; indigenous 7
7. Capsule opening by well-defined minutely transversely
    striate valves; ovary not divided into locules to apex, the
    placenta not continuous with the style . . . . .   9. **Rotala**
  Capsule ± circumscissile or bursting irregularly or if
    opening by valves then not transversely striate; ovary
    completely or incompletely divided into locules 8
8. Flowers solitary; calyx tubular; fruit opening by 2 valves   5. **Lythrum**
  Flowers in axillary cymes or clusters; calyx often bowl-
    shaped or campanulate; fruit ± circumscissile or
    bursting irregularly 9

9. With one or more of the following characters — inflorescences
    surrounded by obvious bracts, style exceeding 3mm.
    (sometimes only in long-styled forms of some species),
    petals larger, 2.5–7 mm. long; also calyx usually longer,
    the tube, 2.5–4.7 mm. long; flowers often heterostylous;
    capsule mostly at first circumscissile around the style-
    base then bursting irregularly   6. **Nesaea** (in part)
    Inflorescences with smaller or minute bracts, style lacking
    or up to 2.5(–3) mm. long, petals smaller, rarely up to 2
    mm. long or absent; calyx mostly under 2.5 mm. long;
    flowers usually not heterostylous (never in *Ammannia*);
    capsule as above or bursting irregularly 10
10. Leaves linear-lanceolate, 2.5–6 × 0.15–0.3 cm.; style 1.5–2
    mm. long; seeds few, ± 6, 1–1.75 mm. long; inflorescences
    dense and sessile; bracts ± 3 mm. long; slender annual
    (T7) . . . .   8. **Hionanthera**
    Without these characters combined; seeds much more
    numerous and smaller   11
11. Calyx-appendages well developed, very distinctly longer
    than the calyx-lobes . . . .   6. **Nesaea** (in part)
    Calyx-appendages less developed or absent but sometimes
    well developed in bud and ± equalling calyx-lobes   7. **Ammannia**

1. WOODFORDIA

Salisb., Parad. Londin., t. 42 (1806); Koehne in E.P. IV. 216: 78, fig. 12 (1903)

Shrubs or small trees. Leaves decussate, subcoriaceous, the lamina finely black-
punctate beneath. Flowers slightly zygomorphic, in axillary cymes, (5–)6-merous. Calyx-
tube tubular or tubular-urceolate; lobes short, ± triangular, alternating with small callous
appendages. Petals small, elliptic to lanceolate-attenuate, shorter or ± longer than the
sepals. Stamens (10–)12, inserted in the tube below the middle, well exserted and often
curved to one side. Ovary sessile, oblong, incompletely 2-locular; style slender, often
curved, exceeding ovary; ovules numerous. Capsule ellipsoid-ovoid, enclosed within the
calyx-tube, thin-walled, breaking irregularly or ± 2-valved. Seeds numerous, compressed
trigonous-ovoid, minute.

Two closely related species both occurring in East Africa and one extending to Asia. None of the
characters used in separating the two is constant but a complete survey of the genus throughout its
range convinced me that it is wise to follow Koehne in recognising two species.

W. fruticosa is cultivated in Nairobi, *Gillett* 16344, Nairobi City Park, 6 Nov. 1964; *Luke*, same place, 8
Sept. 1986 and *Graham Bell* 2, Nairobi, Closeburn, 3 Mar. 1952.

Leaves finely but densely velvety tomentose beneath*; petals
    often exceeding calyx-lobes and more acuminate, sometimes
    ± long-acuminate (T6, Mafia I., also cultivated in Nairobi)   1. *W. fruticosa*
Leaves ± glabrous or with scattered pubescence; petals usually
    shorter than the calyx-lobes and elliptic, not or not so
    acuminate (U1, 3; K2, 3) . . . .   2. *W. uniflora*

1. **W. fruticosa** (*L.*) *Kurz* in Journ. Asiat. Soc. Bengal 40: 56 (1871) & For. Fl. Brit. Burma
1: 518 (1877); Koehne in E.J. 1: 333 (1881) & E.P. IV. 216: 80, t. 12A (1903); T.T.C.L.: 294
(1949); H. Perrier, Fl. Madag. 147, Lythracées: 19, fig 3/5, 6 (1954); Dar in Fl. W. Pak. 78: 6,
fig. 1D–F (1975). Type: ? China, *Linnaean Herbarium* 626/4 (LINN, lecto.)

Shrub or small tree 1.3–5 m. tall, with spreading pubescent or glabrous stems. Leaves
subsessile or shortly petiolate; blades ovate to ovate-lanceolate or lanceolate, 2–14 cm.
long, 1–4 cm. wide, attenuate at the apex, rounded to subcordate at the base, finely but

* A high-power microscope is needed to see this properly; in East Africa the differences are clear.

densely velvety tomentose beneath or in some areas (see note) glabrous. Inflorescences crimson, (1–)2–16-flowered, up to ± 3 cm. long; pedicels 0.4–1 cm. long; bracteoles minute, Calyx-tube 0.9–1.5 cm. long, 2–5 mm. wide, glabrous or puberulous; lobes oblong-ovate. Petals red, lanceolate-acuminate, 3–4.5 mm. long, 0.5–0.8 mm. wide, equalling or longer than calyx-lobes. Stamens 0.3–1.7 cm. long. Ovary 4–6 mm. long; style 0.8–1.5 cm. long. Capsule 0.6–1 cm. long, 2.5–4.5 mm. wide. Fig. 1.

TANZANIA.   Rufiji District: Mafia I., Bweni–Mawalani, 17 Aug. 1937, *Greenway* 5130!
DISTR.   **T** 6; Arabia (see note), Yunnan, Pakistan, India, Burma, Sri Lanka, Thailand, Java, Timor, W. Sumbawa, Madagascar, Comoro Is.; also cultivated in Kenya, Trinidad and Mauritius
HAB.   *Sideroxylon, Mimusops, Terminalia fatraea, Bridelia*, etc., open coastal bushland on coral rock; ± sea-level

SYN.   *Lythrum fruticosum* L., Sp. Pl., ed. 2: 641 (1762)
   Grislea tomentosa Roxb., Pl. Coromandel 1: 29, t. 31 (1795): Bot. Reg. l, t. 30 (1815); Roxb., Fl. Indica 2: 233 (1832), *nom. illegit.* Type as for *Lythrum fruticosum*
   Woodfordia floribunda Salisb., Parad. Londin, t. 42 (1806); Hiern in F.T.A. 2: 481 (1871), pro parte; C.B. Cl. in Fl. Brit. Ind. 2: 572 (1879), pro parte, *nom. illegit.* Type: as for *W. fruticosa*
   Grislea punctata Smith in Rees, Cyclop. 17, n. 2 (1819); Wight & Arn., Prodr. Fl. Pen. Ind. Or.: 308 (1834). Type: E. Indies, *Buchanan* in *Herb. Smith* 659.2 (LINN, holo., microfiche!)
   Lythrum hunteri DC., Prodr. 3: 83 (1828). Type: India, 'Málava', W. Hunter reference in Asiatic Res. 4: 42 (1795)
   Woodfordia tomentosa (Roxb.) Bedd., Fl. Sylv. S. India: cxvii (Anal. Gen.), t. 14, fig. 4 (1872), *nom. illegit.*

NOTE.   The single East African gathering seen from a wild source is clearly identical with the Madagascan populations as might be expected and has the tomentum well developed. In some areas it must be admitted that the two species are not well separated, e.g. in Arabia where the tomentum is less pronounced and the petals certainly not always longer than the calyx-teeth. *Schweinfurth* 681 (Yemen) at least is closer to *W. fruticosa*, confirmed by S.A. Graham (pers. comm.), who has both species in cultivation and who suggests it might be *W. fruticosa* (♀) × *W. uniflora* (♂). She also points out that the shape and size of the calyx-tube of the two taxa usually differ from that of *W. fruticosa*, being longer and more slender. In most areas, e.g. Yunnan, the tomentum is thick and the petals much longer that the calyx-teeth, but some specimens in Assam, Madras, Burma and Thailand have glabrous leaves but usually long petals. Koehne (E.P. IV. 216: 80) has forma *genuina* Kurz, i.e. forma *fruticosa*, with greyish silvery pubescence and forma *punctata* (Smith) Blume with subglabrous leaves. Jex-Blake (Gard. E. Afr., ed. 4: 129 (1957)) mentions *Woodfordia floribunda* in cultivation but it is impossible to tell which of the two species he actually means. In India the flowers are used to produce a yellow or red dye.

2. **W. uniflora** (*A. Rich.*) *Koehne* in E.J. 1:334 (1881) & E.P. IV. 216: 80, fig 12B (1903); F.P.S. 1:143 (1950); F.W.T.A., ed. 2,1:163 (1954); E.P.A.:610 (1959); K.T.S.: 258 (1961); Burger, Fam. Fl. Pl. Eth.: 184, fig. 46/2 (1967); M.G. Gilbert in Fl. Eth., ined. Types: Ethiopia, Tigray, Mai Gouagoua, *Quartin-Dillon* (P, syn.) & Djeladjeranne, *Schimper* 1906 (P, syn, K, isosyn.!)*

Slender often scrambling shrub or small tree 0.8–5 m. tall; stems puberulous; bark light brown on older stems breaking into fibrous strips. Leaves shortly petiolate, opposite or sometimes in 3's (fide F.P.S.), lanceolate, (1–)4–7(–12.5) cm. long, 0.4–3.5 cm. wide, acuminate at the apex, cuneate to rounded or subcordate at the base, glabrous, puberulous or pubescent on the venation beneath but not densely tomentose. Inflorescence 1–6(–many)-flowered, up to 3 cm. long; pedicels 3–8 mm. long; bracteoles 2–4 mm. long. Calyx-tube red or orange-red, yellowish beneath, 1–1.2 cm. long, 3–4 mm. wide, glandular-pubescent or puberulous; lobes triangular, (1–)1.5–2(–3) mm. long, acute. Petals pink, red or whitish, elliptic, 0.5–2 mm. long, 0.2–0.3(–0.4) mm. wide, shorter than or equalling or rarely longer than the calyx-lobes, sometimes ciliate. Stamens 1.3–1.6 cm. long. Ovary 5–6 mm. long; style ± 1 cm. long. Capsule ± 8 mm. long, 3–4.5 mm. wide.

UGANDA.   Karamoja District: Moroto, Kasumeri Estate, Jan. 1972, *J. Wilson* 2157! & Napak, near Iriri, Nov. 1941, *Dale* U195!; Mbale District: NE. Elgon, Tulel valley, Sept. 1959, *Tweedie* 1901!
KENYA.   W. Suk District: lower slopes of Suk Hills, Oct. 1937, *Gardner* in *F.D.* 3717!; Baringo District: Marigat–Kabarnet road, Sept. 1960, *Dale* K1091! & above Bartabwa on road to Bartolimo, 16 Oct. 1988, *Luke* in *B.F.F.P.* 218!
DISTR.   **U** 1, 3; **K** 2, 3; N. Nigeria, Cameroon, Sudan and Ethiopia; also Arabia

* Gilbert has written up three sheets of *Schimper* 243 (Ethiopia, Schahagenna) (K) as isosyntypes and treats *Schimper* 1906 as a syntype of *Crislea multiflora* A. Rich, A sheet at GOET has label "243, 1906 in Valle Schahagenni."

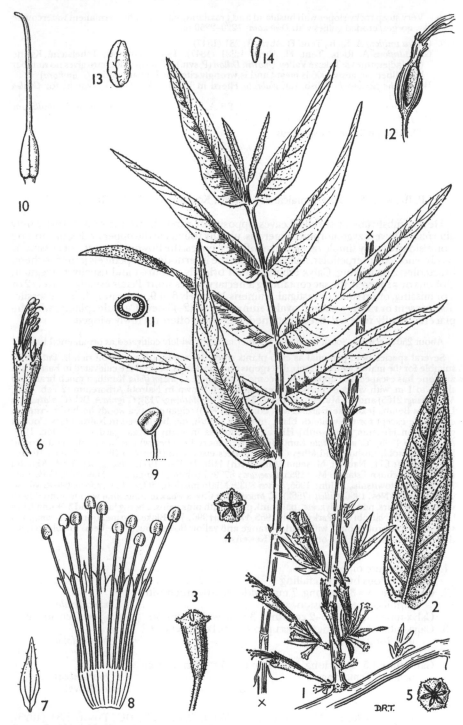

FIG. 1. *WOODFORDIA FRUTICOSA*—1, habit, × 1; 2, leaf undersurface, × 1; 3–5, bud, two views, showing varying number of sepals, × 2; 6, flower, × 2; 7, petal, × 6; 8, flower opened out, gynoecium removed, × 3; 9, anther, × 6; 10, gynoecium, × 3; 11, T.S. of ovary, × 6; 12, fruit, × 2; 13, ripe fruit, × 2; 14, seed, × 12. All from *Greenway* 5130. Drawn by D.R. Thompson.

HAB.   Very steep rocky slopes with bushland and grassland; seasonal and permanent watercourse
banks, gorges, eroded gullies with *Dodonaea*; 1220–2250 m.

SYN.   *Grislea uniflora* A. Rich., Tent. Fl. Abyss. 1: 281 (1847)
        G. multiflora A. Rich., Tent. Fl. Abyss. 1:281 (1847). Types: Ethiopia, Tchelatché Kanné
        (Djeladjeranne) & Takazé Valley, *Quartin Dillon* (P, syn.) & *Schimper* [Richard gives no number
        but Gilbert suggests 1906 is meant and is wrongly cited by Richard under *G. uniflora*]
        Woodfordia floribunda Salisb. var. *glabrata* Hiern in F.T.A. 2:481 (1871). Type as for *Grislea
        uniflora*
        W. floribunda Salisb. var. *tomentosa* Hiern in F.T.A. 2: 481 (1871). Type as for *Grislea multiflora*
        [*W. fruticosa* sensu Collenette, Fl. Saudi Arabia: 357 (1985) *non* (L.) Kurz *fide* S.A. Graham]

NOTE.   Possibly pollinated by sun-birds.

2. CUPHEA

P. Browne, Nat. Hist. Jamaica: 216 (1756); Koehne in E.P. IV. 216: 80 (1903)

Herbs, subshrubs or shrubs. Leaves opposite or rarely in whorls of 3–5, very rarely
alternate. Flowers zygomorphic, 6-merous, usually solitary at the upper nodes, or forming
± one-sided usually simple raceme-like inflorescences, the flowers alternate, decussate or
rarely whorled, interpetiolar, axillary or from the internodes, the pedicels joined to them;
bracteoles 2 or lacking. Calyx tubular, often brightly coloured and usually ± irregular,
gibbous or ± spurred, the sac containing a nectary; lobes short. Petals usually 6, rarely 2 or
4 or missing, often small or vestigial. Stamens 11, rarely 6 or 9, very rarely 4. Ovary sessile;
disc present or rarely ± absent; ovules numerous or (2–)3–few. Capsule splitting with the
placenta finally reflexed. Seeds flattened, biconvex, often narrowly winged.

About 250 species in South and Central America many widely cultivated as ornamental herbs.

Several species hold potential as crop plants since the seeds contain an oil rich in fatty acids
suitable for the manufacture of soap, detergents and lubricants. Several are cultivated in East Africa
and one has escaped and apparently become established. *C. hyssopifolia* Kunth, a much branched
shrub to 1 m. with violet or purple flowers, has been grown in Nairobi Arboretum (2 Feb. 1982,
Mwangangi 2165) and the Museum garden (23 May 1970, *Mathenge* 713); *C. ignea* A. DC. (*C. platycentra*
Lem. *non* Benth., Jex-Blake, Gard. E. Afr., ed. 4: 75 (1957)) (cigar plant), a woody herb with crimson
calyx-tube except for the white or black and grey mouth, has been used in Nairobi City Council
gardens on Museum [Ainsworth] Hill and doubtless in many private gardens (16 Dec. 1971,
Mwangangi 1903); *C. micropetala* Kunth (dealt with more fully below) has been cultivated around
Hoima, Nairobi, Lushoto and Mbeya and doubtless elsewhere (Bunyoro District, Hoima, 16 Oct.
1970, *Katende* 671; Nairobi, Museum [Ainsworth] Hill, 16 Dec. 1971, *Mwangangi* 1904; Kiambu
District, Closeburn Estate, 6 May 1953, *Greenway* 8779; Lushoto, Mangula, 18 May 1979, *Chilangola*
131; Lushoto township, 15 June 1965, *Semsei* 3939; Mlalo mission, 29 Jan. 1987, *Kisena* 646 & Mbeya
guest house, 3 Nov. 1966, *Gillett* 17562); *C. procumbens* Cav. a weak to somewhat woody annual herb
with purple hairs, purple calyx, etc., pale pink petals with pink veins, 2 being large and 3–4 small, has
been grown in Nairobi City Park (29 July 1953, *Verdcourt* 996). Jex-Blake also mentions a *C. eximia* as a
straggling subshrubby plant with scarlet-orange and yellow flowers but I have been unable to trace
this name in either Koehne or the Index Kewensis.

1. Petals minute or lacking   2
   Petals obvious but soon falling   3
2. Calyx-tube to ± 3 cm. long, 7 mm. wide, yellow-green with
     vermilion-red suffusion   .   .   .   .   *C. micropetala*
   Calyx-tube smaller, 1.5–2 cm. long, 3 mm. wide, crimson   *C. ignea* (see above)
3. Calyx-tube ± 5 mm. long; leaves up to 1.5 × 0.4 cm.; petals ± 2
     mm. long.   .   .   .   .   *C. hyssopifolia*
                                                                                (see above)

   Calyx-tube 1.5–2 cm. long; leaves up to 3.5 × 1.5 cm.; large
     petals ± 1 cm. long.   .   .   .   .   *C. procumbens*
                                                                                (see above)

C. micropetala *Kunth*, Nov. Gen. Sp. 6: 209, t. 551 (1824); DC., Prodr. 3:84 (1828);
Koehne in Fl. Bras. 13: 332 (1877) & in E.J. 2: 400 (1882) & in E.P. IV. 216: 161, fig. 22D
(1903); Bailey, St. Cycl. Hort., rev. ed. (1): 913 (1939); Coode, Fl. Masc. 95, Lythracées:
12(1990), M.G. Gilbert in Fl. Eth., ined. Type: plant grown in "horto botanico Mexicano"
(P?, holo.)

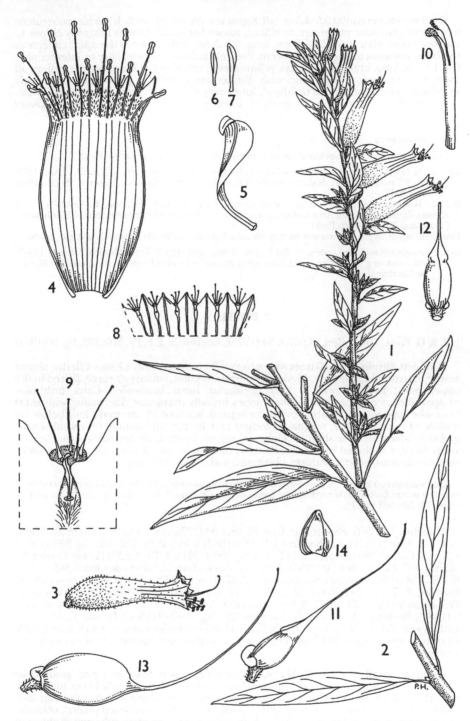

FIG. 2. *CUPHEA MICROPETALA*—1, habit, × ⅔; 2, mature (opposite) leaves, × ⅔; 3, flower, × 1; 4, flower, opened out, × 2; 5–7, petals, × 10; 8, top part of corolla, × 2; 9, petal and part of corolla-rim, much enlarged; 10, stigma and part of style, much enlarged; 11, gynoecium, × 2; 12, ovary (dorsal view), × 2; 13, fruit, × 2; 14, seed, × 6. 1–12, from *Verdcourt* 3048; 13, 14, from *Pringle* 11450 (Mexico). Drawn by Pat Halliday.

Erect woody perennial 0.5–1.5 m. tall. Stems usually bright reddish purple, puberulous to densely glandular pubescent or villous, somewhat viscid. Leaves narrowly elliptic to narrowly lanceolate, 0.5–12(–16) cm. long, 0.8–3 cm. wide, acute at the apex, cuneate at the base, glabrous to shortly pubescent. Pedicels axillary, extra-axillary or interpetiolar, 0.3–1.1 cm. long, bracteolate. Calyx yellow-green, vermilion at base, later vermilion all over, ± 3 cm. long, 7 mm. wide, with distinct basal swelling, puberulous to densely glandular-pubescent, villous or hispid; lobes very short, alternating with setose sinus-horns. Petals white or yellowish, vestigial. Stamens long-exserted. Disc excentric. Ovary saccate near apex. Style exserted 1–1.5 cm. Capsule many-seeded.

var. **micropetala**

Leaves ± glabrous. Calyx puberulous and sometimes shortly sparsely pilose. Fig. 2.

KENYA. N. Nyeri District: point at where Mweiga D 449 road crosses R. Amboni (Honi), 12 Aug. 1977, *Hooper & Townsend* 1691!; Nairobi, banks of Nairobi R. below National Museum, 26 Jan. 1961, *Verdcourt* 3048!

DISTR. **K** 4; native of Mexico, cultivated in Europe, Egypt, Rwanda, Ethiopia, E. Africa (see note above), Zimbabwe, South Africa, Madagascar, Java, etc.; certainly long established in South Africa (Transvaal, Krugersdorp Park)

HAB. Banks of rivers and streams in very disturbed ground formerly *Croton, Cordia*, etc., forest

SYN. *C. eminens* Planch. & Linden, Fl. des Serres 10: 69, t. 994 (1854). Type: Mexico, *Andrieux* 136 (P, syn.), drawing by Linden (syn.), specimens grown by Linden from seeds collected in Mexico by *Ghiesbreght* (syn.)

3. **PEMPHIS**

J. R. & G. Forst., Char. Gen. Pl.: 67, t. 34 (1776); Koehne in E.P. IV. 216: 185, fig. 30(1903)

Shrublets, shrubs or small trees with silvery-silky indumentum. Leaves ± fleshy, almost sessile, narrowed to the base. Flowers 6-merous, regular, solitary or rarely paired in the upper axils; pedicels with 2 very deciduous basal bracteoles. Calyx turbinate-campanulate, ± coriaceous, 12-ribbed; lobes broadly triangular, alternating with short horn-like appendages. Petals obovate, corrugated. Stamens 12, inserted a little below the middle of the calyx-tube, ± equal or longer and shorter alternating. Ovary subsessile, globose, 3-locular; style short or elongate, stigma 2-lobed; ovules numerous. Capsule subglobose, ± included in the corolla-tube, ultimately almost 1-locular, circumscissile. Seeds numerous, erect, imbricate, obcuneate with thickened wing.

One widespread littoral species; a species from low mountains in SW. Madagascar is now referred to a new genus *Koehneria* (see exhaustive paper by S.A. Graham, Tobe & Baas in Ann. Missouri Bot. Gard. 73: 788–809 (1987)).

P. acidula *J. R. & G. Forst.*, Char. Gen. Pl.:68, t. 34 (1776); DC., Prodr. 3: 89 (1828); Hiern in F.T.A. 2: 482 (1871); Koehne in E.J. 3: 133 (1882) & in E.P. IV. 216: 185, fig. 30B (1903); Gilg in P.O.A. C: 285 (1895); V.E. 3(2): 651, fig. 285 (1921); T.T.C.L.: 294 (1949); Perrier, Fl. Madag. 147, Lythracées: 20 (1954); A. Fernandes & Diniz in Garcia de Orta 4: 389 (1956); K.T.S.: 258 (1961); Gomes e Sousa, Dendrol. Moçamb. 2: 555, t. 166 (1967); Lewis in Proc. Roy. Soc. B 178: 79–94 (1971) & 188: 247–256 (1975) (heterostyly); Tseng Chieng Huang in Fl. Taiwan 3:819, t. 827 (1977); A. Fernandes in F.Z. 4: 281, t. 68 (1978) & in Fl. Moçamb. 73: 6 (1980); Fosberg & Renvoize, Fl. Aldabra: 132, fig. 19/7–9 (1980); Coode, Fl. Masc. 95: Lythracées: 6, t. 3 (1990). Type: not indicated but Fernandes states syntypes from Pacific Islands (a specimen at BM from Teootea (= Takaroa) and Savage I. (= Niue) may be a syntype)

Shrub or small tree 0.6–8(–11) m. tall, usually much branched and gnarled or occasionally in poor coral soil situations a dwarf creeping shrublet only 15 cm. high; bark dark grey, rough, reticulate and flaking; nodes rather thickened with conspicuous leaf-scars. Leaves leathery to distinctly fleshy (particularly in very exposed plants), obovate-oblanceolate to linear-lanceolate, 0.8–3.5 cm. long, 0.3–1.3 cm. wide, subacute or obtuse at the apex, narrowed to the base, 1-nerved. Flowers sweet-scented, heterostylous; pedicels 0.5–1.3 cm. long; bracteoles ± 4 mm. long. Calyx greenish, lobes sometimes with reddish edges. Petals white or pink, 5 mm. long, 4 mm. wide, very soon falling. Short-styled flowers: stamens subbiseriate, alternately unequal, the longer slightly exceeding the

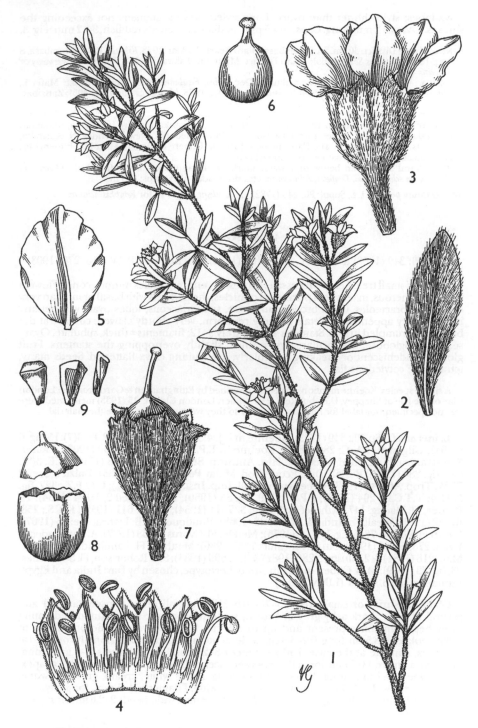

FIG. 3. *PEMPHIS ACIDULA*—1, habit, × ⅔; 2, leaf, × 2; 3, flower, × 6; 4, calyx, × 6; 5, petal, × 8; 6, gynoecium, × 8; 7, fruit, × 4; 8, fruit, dehiscing with seeds, × 4. 1, 7, 8, from *Pedrogão* 5199; 2–6, from *Torre & Paiva* 12123. Drawn by Victoria Goaman. Reproduced with permission from Flora Zambesiaca.

calyx-lobes; style shorter than ovary. Long-styled flowers: stamens not exceeding the sinuses. Ovary ± 2 mm. long. Capsule ± 5 mm. in diameter. Seeds reddish, 3 × 2 mm. Fig. 3.

KENYA. Kwale District: 30 km. S. of Mombasa, Diani Beach, 26 Mar. 1973, *Kibuwa* 1204!; Mombasa, 8 Aug. 1965, *Williams Sangai* 865!; Lamu District: Manda I., Takwa Creek, 8 Feb. 1956, *Greenway & Rawlins* 8873!
TANZANIA. Tanga District: Ras Nyamaku, 23 Dec. 1956, *Faulkner* 1928!; Rufiji District: Mafia I., Kanga, 11 Aug. 1937, *Greenway* 5046!; Lindi Bay, 22 Oct. 1978, *Magogo & Rose Innes* 405!; Zanzibar, I. Mazizini [Massazine], 21 Nov. 1959, *Faulkner* 2406!
DISTR. **K** 7; **T** 3, 6, 8; **Z**; **P**; Mozambique, Madagascar, Europa I., Mascarenes, Seychelles, Aldabra; also China (Hupeh), Taiwan, Ryuku Is., India (Madras), Sri Lanka, Maldive Is., Burma, Andaman Is., Malaya, Thailand, Malesia (to New Guinea), New Caledonia, New Hebrides, N. Australia, Pacific Is. from Mariana Is. and Caroline Is. to Kiribati [Gilbert] & Tuvalu [Ellice] Is., Phoenix Is., Fiji, Tonga, Society Is., Tuamotu Is. and Pitcairn I.
HAB. Littoral at or even below high-water mark, coral-rag thicket with *Mimusops*, *Sideroxylon*, *Combretum schumannii*, *Drypetes*, *Adansonia*, etc., drier sides of swamps; 0–5 m.

SYN. *Lythrum pemphis* L.f., Suppl. Pl.: 249 (1782), *nom. illegit.* Type: as for *Pemphis acidula*

4. LAWSONIA

L., Sp. Pl: 349 (1753) & Gen. Pl., ed. 5: 166 (1754); Koehne in E.P. IV. 216: 270 (1903)

Shrub or small tree. Leaves decussate, subsessile, entire; stipules minute, conic. Flowers regular, 4-merous, in terminal pyramidal panicles, the pedicels with basal or median very deciduous bracteoles. Calyx broadly turbinate, subcoriaceous; lobes ovate without any intermediary appendages. Petals broadly reniform, very shortly clawed, cordate at the base, very crumpled in bud. Stamens 4, 8 or rarely 9–12; filaments ± thick, subulate. Ovary sessile, subglobose, 2–4-locular; style ± thick, slightly overtopping the stamens. Fruit globose, indehiscent or breaking irregularly, the fruiting calyx flattened. Seeds many, tetragonal; cotyledons flat.

A single species. Koehne mentions a species described by Ettinghausen & Gard. from the London Clay of the Isle of Sheppey, U.K., but Reid & Chandler, London Clay Fl.: 15 (1933) mentioned there was no specimen nor label for *Lawsonia europaea* so they were unable to revise the material.

L. inermis *L.*, Sp. Pl.: 349 (1753); Koehne in E.J. 4:36 (1883) & in E. & P. Pf. 3(7):15, fig. 6 (1891); Gilg in P.O.A. C: 286 (1895); Koehne in E.P. IV. 216:270, fig. 59 (1903); Laufer, Sino-Iranica, Publ. Field. Mus. Nat. Hist. Anthrop. Ser. 15(3): 334–338 (1919); V.E. 3(2): 655, fig. 288 (1921); Burkill, Dict. Econ. Prod. Malay Pen. (2): 1323 (1935); Dalziel, Useful Pl. W. Trop. Afr.: 40 (1937); Greenway in Bull. Imp. Inst. 39: 236 (1941); U.O.P.Z.: 327, fig. (1949); T.T.C.L.: 294 (1949); F.P.S. 1:139, fig. 85 (1950); F.W.T.A., ed.2, 1:164 (1954); H. Perrier, Fl. Madag. 147, Lythracées: 23, fig. 3/7–11 (1954); E.P.A.: 611 (1959); K.T.S.: 258, fig. 52 (1961); Wealth of India 6: 47–50, t. 3 (1962); Boutique, F.C.B. Lythraceae: 25 (1967); Dar in Fl. W. Pak. 78: 5, fig. 1/A–C (1975); Meikle, Fl. Cyprus 1: 668 (1977); A. Fernandes in F.Z. 4: 278, fig. 66 (1978) & in Fl. Moçamb. 73:3 (1980); Matthew, Fl. Tamilnadu Carnatic 1, Mat.: 218 (1981) & 2, Illustr., t. 280 (1982) & 3, 1: 609 (1983); S.A. Robertson, Fl. Seychelles: 93, fig. 73 (1989); M.G. Gilbert in Fl. Eth., ined. Lectotype, chosen by Dar: India and Egypt; Linnaean Herb. 496/1 (LINN, lecto.)

Glabrous shrub or small tree 1.5–7(–10.5–?12*) m. tall often densely tangled and several m. wide; bark ashy grey or brown, smooth or striate; spines (modified young branches) sometimes present and up to 3.5 cm. long. Leaves elliptic or oblong to oblanceolate, 1–8.5 cm. long, 0.2–3.8 cm. wide, acute, apiculate or occasionally rounded at the apex, cuneate at the base. Inflorescences 3–25 cm. long; pedicels 2–4 mm. long; bracteoles linear, ± 0.5 mm. long; flowers sweet-scented; buds sometimes reddish at apex. Calyx-tube 1–1.7 mm. long; lobes 2–3 mm. long. Petals white, cream, greenish white, green or yellow, 1.5–2.5(–4) mm. long, 4–5 mm. wide. Filaments 4–5 mm. long. Style ± 3 mm. long. Fruits purplish green, 4–6 mm. long, 4.5–8 mm. wide. Seeds 2–2.6 mm. long. Fig. 4.

UGANDA. Karamoja District: Kidepo [Kidapo] R., Apr. 1960, *J. Wilson* 909!

---

* not exceeding 40 feet *fide* Battiscombe.

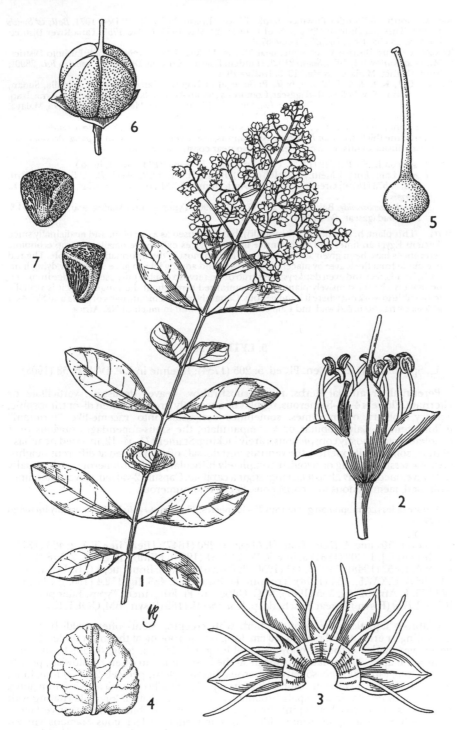

FIG. 4. *LAWSONIA INERMIS*—**1**, flowering branch, × ⅔; **2**, flower, × 8; **3**, calyx, opened to show filaments, × 6; **4**, petal, × 8; **5**, pistil, × 8; **6**, fruit, × 4; **7**, seeds, two views, × 6. 1–5, from *de Carvalho* 1884–1885; 6, 7, from *Mendonça* 2351. Drawn by Victoria Goaman. Reproduced with permission from Flora Zambesiaca.

KENYA. Northern Frontier Province: bank of Uaso (Ewaso) Nyiro R., 19 Dec. 1971, *Bally & Smith* 14693!; Turkana District: 128 km. S. of Lodwar, 21 May 1953, *Padwa* 175!; Tana River District: Kurawa, 28 Oct. 1961, *Polhill & Paulo* 673!
TANZANIA. Pare District: W. of Same, near Marua, 11 Aug. 1965, *Leippert* 6068!; Lushoto District: Mkomazi valley, July 1955, *Semsei* 2132!; Handeni District: Kangata, Dec. 1949, *Semsei* in *F.D.* 2890!; Rufiji District: Mafia I., 28 Mar. 1933, *Wallace* 793!
DISTR. **U** 1; **K** 1, 2, 4, 7; **T** 3, 5, 6; **Z**; **P**; Senegal to Nigeria, Zaire, Ethiopia, Somalia, Sudan, Mozambique, Seychelles, Madagascar, Comoro Is., South Africa (Natal); also Palestine, Syria, Iraq, Arabia, Egypt, Libya, China, Pakistan (wild in Baluchistan), India, Maldive Is., Sri Lanka, Malaya, Indochina, Malesia to New Guinea and Australia (see note)
HAB. Temporarily flooded rocky river courses in dry country with *Acacia, Salvadora, Capparidaceae*, etc., riverine thicket, *Cordia-Hyphaene* associations, also riverine forest with *Populus, Acacia, Ficus, Kigelia*, hillsides, cliffs, rock-crevices, etc., also near coast; ± 2–1350 m.
SYN. *L. spinosa* L., Sp. Pl.: 349 (1753). Type: Sri Lanka, *Hermann* (BM-HERM, lecto.)
　L. alba Lam., Encycl. Méth. Bot. 3:106 (1789); Boiss., Fl. Orient. 2: 744 (1872); C.B. Cl. in Fl. Brit. Ind. 2: 573 (1879); Greenway in E. Afr. Agric. Journ. 6: 131 (1941) , *nom. illegit*. Type as for *L. inermis*
　Rotantha combretoides Bak. in J.L.S. 25: 317, t. 51 (1890). Types: central Madagascar, *Baron* 2194 & NW. Madagascar, *Baron* 5032, 5169 (K, syn.!)
NOTE. This plant, henna (mhina (Kiswahili)), has been used as a dye plant and medicinally since Ancient Egyptian times and an extensive literature gives exhaustive details — some economic references have been given above which will lead to most of the information available. The red dye made from the leaves by macerating in water and adding lemon juice is used for dying hair, nails, hands, feet and even donkeys; the shrub often serves a double purpose as a hedge in coastal plantations. Now extensively planted and naturalised throughout the range given it is virtually impossible to work out where it is indigenous. Certainly in Baluchistan, parts of India and SW. Asia it does or has occurred wild and I am certain it is also wild in much of NE. Africa.

5. LYTHRUM

L., Sp. Pl.: 446 (1753); Gen. Pl., ed. 5: 205 (1754); Koehne in E.P. IV. 216: 58 (1903)

Perennial or annual herbs, rarely subshrubby. Leaves decussate, verticillate or alternate. Flowers 4–6(–8)-merous, regular or obscurely irregular, often di- or trimorphic, axillary, solitary or in cymes, sometimes congested into raceme-like "terminal" inflorescences. Calyx tubular or ± campanulate, the sinus-appendages obvious or ± obsolete. Petals mostly conspicuous, rarely lacking. Stamens (1–)4–12, inserted near base of tube, some or all exserted, the ventrals and dorsals often inserted at different heights. Ovary ± sessile, oblong or ovoid, incompletely bilocular; ovules numerous; style usually well developed, rarely almost lacking; stigma capitate. Capsule 2-valved; valves sometimes ± bilobed, membranous to ± coriaceous. Seeds 8 to numerous.

A more or less cosmopolitan genus with 35–40 species. The genus *Peplis* L. is now usually included in *Lythrum*.

L. rotundifolium A. *Rich.*, Tent. Fl. Abyss. 1: 280 (1847); Hiern in F.T.A. 2: 465 (1871); Koehne in E.J. 1: 308 (1881) & in E.P. IV. 216: 61 (1903); Mildbr. in Z.A.E. 2: 578 (1913); F.P.N.A. 1: 657 (1948); F.P.S. 1: 141 (1950); E.P.A.: 610 (1958); Boutique, F.C.B., Lythraceae: 13 (1967); U.K.W.F.: 154 (1974); Troupin, Fl. Rwanda 2: 483, fig. 152.4 (1983); Blundell, Wild Fl. E. Afr.: 52, fig. 565 (1987); M.G. Gilbert in Fl. Eth., ined. Types: Ethiopia, Shoa [Choa], *Petit* (P, syn.) & near Enschedcap, *Schimper* II.1169 (P, syn., BM, GOET, K, isosyn.!)

Glabrous aquatic or semiaquatic herb with creeping stems often reddish, having ascending or erect apical parts, 5–60 cm. long or tall, rooting at the lower nodes. Leaves decussate; blades ± round, obovate or oblong, 0.5–2.5 cm. long, 0.35–1.1 cm. wide, rounded at apex, subcordate or truncate at the base; venation often ± yellowish; petiole 0.5–3.5 mm. long. Flowers solitary, 4-merous, heterostylous; pedicels 2–5(–7) mm. long; bracteoles linear, 1–1.5 mm. long. Calyx often pink, broadly cylindrical, 3.5–5 mm. long, 2.5–3 mm. wide, 8-veined; sepals broadly triangular, 0.8–1 mm. long, alternating with narrower sinus-horns. Petals pink, blue-purple, magenta or blue with darker venation, obovate, 6–8 mm. long, 4.5–5 mm. wide, almost immediately deciduous. Stamens 8 in two unequal whorls; short-styled flowers with stamens 5–7.5 mm. long; long-styled flowers with stamens not exceeding the sepals; anthers blue. Ovary oblong-ovoid, 2–3.5 mm. long; style up to 1 mm. long in short-styled flowers, 4–4.5 mm. long in long-styled flowers; stigma capitate. Capsule ellipsoid, 5–6 mm. long. Fig. 5.

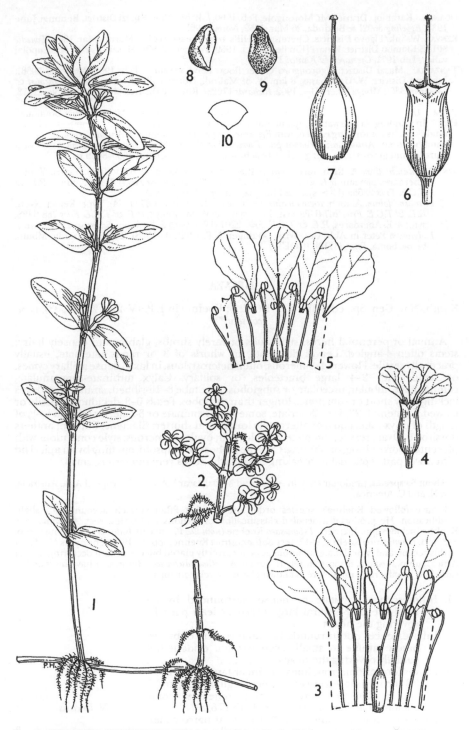

FIG. 5. *LYTHRUM ROTUNDIFOLIUM*—**1**, **2**, habit, × ⅔; **3**, flower, opened out, × 4; **4**, flower, × 2⅖; **5**, flower, opened out, × 4; **6**, fruit, × 4; **7**, fruit, calyx removed, × 4; **8, 9**, seeds, two views, × 12; **10**, T.S. seed, diagrammatic. 1, from *Purseglove* 3138; 2, from *Hepper & Field* 4876; 3, from *Eggeling* 3692; 4–10, from *Greenway* 14984. Drawn by Pat Halliday.

UGANDA. Karamoja District: Mt. Morongole, Feb. 1960, *J. Wilson* 825!; Kigezi District: Behungi, June 1938, *Eggeling* 3692! & Bukinda, 26 Mar. 1952, *Norman* 96!

KENYA. W. Suk/Elgeyo Districts: Cherangani Hills, upper reaches of R. Marun, Mar. 1965, *Tweedie* 3018!; Kiambu District: Kinari [Kinale], 8 Oct. 1959, *Verdcourt* 2493!; Masai District: Nasampolai valley, Feb 1971, *Greenway & Kanuri* 14984!

TANZANIA. Masai District: Ngorongoro Crater floor, Lerai stream, 2 Feb. 1962, *Newbould* 5968!; Lushoto District: W. Usambaras, 8 km. W. of Malindi, Ndamanyilu, 30 Aug. 1950, *Verdcourt & Greenway* 336!; Mbeya Mt., 13 Dec. 1962, *Richards* 17023!; Rungwe District: Mwakaleli, 13 Nov. 1913, *Stolz* 2290!

DISTR. U 1, 2; K 2–6; T 2, 3, 7; E. Zaire, Rwanda, Ethiopia, S. Sudan and Malawi* (fide Boutique)

HAB. In standing water, swamps, upland water-meadows, streamsides, etc., often completely filling muddy ditches or forming carpets, with *Erica* etc., Cyperaceae swamps in *Arundinaria alpina* zone, bamboo forest, *Arundinaria, Podocarpus, Pygeum, Lobelia, Allophylus* forest, *Artemisia* scrub with aquatics and grasses, also shingly river beaches in plateau grassland; 1650–3050 m.

SYN. *L. rotundifolium* A. Rich. forma *minus* R.E. Fries in N.B.G.B. 8: 687 (1924). Types: Kenya, Aberdares plateau, *R.E. & T.C.E. Fries* 2365 (UPS, syn., K, isosyn.!) & Mt. Sattima, *R.E. & T.C.E. Fries* 2365a (UPS, syn.) & Ethiopia, *Schimper* 116 (S, syn.)

L. rotundifolium A. Rich. forma *majus* R.E. Fries in N.B.G.B. 8: 687 (1924). Type: Kenya, Nyeri, *R.E. & T.C.E. Fries* 207 (UPS, syn., K, isosyn.!) & W. Mt. Kenya, *R.E. & T.C.E. Fries* 758 (UPS, syn.) & E. Aberdares, *R.E. & T.C.E. Fries* 2201 (UPS, syn.) & Ethiopia, *Schimper* 225 (S, syn.)

L. lyratum Peter in Abh. Ges. Wiss. Göttingen N.F. 13(2): 85(1928). Type: Tanzania, Mbulu, below boma, *Peter* 43710b (GOET?)

6. NESAEA

Kunth, Nov. Gen. Sp., ed. folio, 6: 151 (1823); Koehne in E.P. IV. 216: 221 (1903), *nom. conserv.***

Annual or perennial herbs, subshrubs or rarely shrubs, glabrous to densely hairy; stems often 4-angled. Leaves decussate, in whorls of 3 or rarely alternate, usually practically sessile. Flowers 4–8-merous, often heterostylous, in lax to dense axillary cymes, or heads with 2–4 large bracteoles, or solitary. Calyx turbinate-campanulate, campanulate, tubular, urceolate or subglobose, the tube 8–16-ribbed; sinus-appendages lacking, very short or sometimes longer than the lobes. Petals 0–8, deciduous, sessile or clawed. Stamens 4–23, 1- or 2- seriate, sometimes geminate or 3-nate, inserted at $\frac{1}{6}$–$\frac{1}{2}$ of length of calyx-tube, equal or alternately longer and shorter; filaments filiform; anthers dorsifixed. Ovary sessile, completely 2–5-locular; ovules numerous; style continuous with placenta, short or long, or wanting. Capsule globose or ellipsoid, opening by an apical lid with lower part septicidal or breaking irregularly. Seeds very numerous, small.

About 55 species, predominantly in Africa and Madagascar but some in tropical Asia, Australia, and N. and C. America.

I have followed Koehne's species order used in his Pflanzenreich account with slight modification. He produced a detailed classification of sections and series: sect. *Ammaniastrum* Koehne includes spp. 1, 2; Sect. *Typonesaea* Koehne (*nom illegit.* since includes type of genus), spp. 3–8; sect. *Heimiastrum* Koehne, 9–17; sect. *Salicariastrum* Koehne, spp. 18, 19. Koehne puts *N. schinzii* in this section and typical material would appear correctly placed but the variant occurring in East Africa, subsp. *subalata*, is more closely related to *N. kilimandscharica*. However, I have hesitated to raise it to specific rank until more is known about the relationships.

1. Inflorescences capitate or dense, surrounded by an involucre of 2–several large bracts at least partially covering the heads 2

Inflorescences not surrounded by an involucre of bracts; bracts minute to small, even if quite evident not partially covering the heads 5

2. Leaves long and narrow, linear to linear-lanceolate, 2–8 × 0.3–1 cm.; inflorescences very dense, globose, up to 1.5 cm. in diameter, terminal and sometimes some additional axillary; strictly erect herb (**T6, 8**) . . . 9. *N. linearis*

Leaves not linear to linear-lanceolate, or if narrow then much smaller and inflorescences small 3

* Not given in F.Z. and no material at Kew.
** Usually dated from Juss., Gen. Pl.: 332 (1789) but there merely mentioned under *Lythrum*.

3. Usually larger mostly perennial herbs (0.08–)0.2–1.2 m.
   long or tall with leaves 1–5.5 × 0.4–2.5 cm.; peduncles
   0.7–5.5 cm. long; style (3.5–)*6–8 mm. long   . . .   6. *N. radicans*
   Smaller annuals or ephemerals 2.5–35 cm. tall; leaves 0.5–
   3.5 × 0.5–1 cm.; peduncles (0–)0.5–2.7 cm. long; style
   1.5–3.5 mm. long   4
4. Leaves narrowed to the base   . . . .   7. *N. erecta*
   Leaves truncate to cordate at the base   . . . . . .   8. *N. cordata*
5. Style almost obsolete or very short, together with stigma
   0.3–0.8 mm. long; stamens 4(–6), very short, not
   exserted or only reaching tips of calyx-lobes; petals
   absent or sometimes 1 minute in *N. burttii* or very rarely
   present in *N. aspera*   6
   Style more developed (if scarcely 0.5 mm. long, in *N.
   crassicaulis*, then petals nearly always present); stamens
   4–16, usually much longer; petals present, mostly
   well developed   7
6. Inflorescences sessile and condensed with both peduncles
   and pedicels very short or obsolete (T5)   . . . .   1. *N. burttii*
   Inflorescences subsessile but laxer with at least pedicels
   evident (T4)   . . . .   2. *N. aspera*
7. Leaves narrowed to the base, narrowly elliptic, 0.5–6 ×
   0.2–2 cm., obtuse to subacute at the apex; plant aquatic
   to subaquatic, suberect to prostrate, often rooting at
   lower nodes   8
   Leaves truncate, subcordate or distinctly auriculate at the
   base, usually thin, mostly more attenuate at the apex or
   if base ± narrowed or subtruncate or rounded these
   leaves smaller and narrower, often ± stiff   10
8. Style 1.25–4 mm. long (1.25, 2.5 or 4 according to form);
   stamens ± 4.5 mm. long in short-styled flowers, ⅔
   exserted even in long-styled flowers; bracteoles
   minute   . . . .   4. *N. pedicellata*
   Style 0.5–1 mm. long; stamens very short, included, or if
   style 5 mm. long (in long-styled *N. triflora*, not seen in
   East Africa) then bracteoles evident, ± 3 mm. long   9
9. Flower clusters dense, pedunculate, the peduncles 1–2.5(–
   4) cm. long, much exceeding the pedicels; bracteoles
   evident, 3–3.5 mm. long   . . . .   5. *N. triflora*
   Flower clusters laxer, the pedicels longer than the short
   peduncles; bracteoles minute or obsolete   . . . .   3. *N. crassicaulis*
10. Leaves large, 4–11 × 0.8–3 cm., the inframarginal nerve
    usually very distinct; inflorescences essentially sessile,
    dense, ± 20-flowered (T 6, 8)   . . . .   18. *N. maxima*
    Leaves smaller or if almost as large as lower limits above
    then inflorescences not as above   11
11. Inflorescences usually 2 per axil or inflorescences
    branched below, lax, peduncles well developed, 0.9–
    1.5 cm. long; leaves 1.5–4.8 × 0.5–1.2 cm., auriculate at
    base (T 2)   . . . .   17. *N. volkensii*
    Inflorescences single, 1–several-flowered dichasia not
    branched below   12
12. Bracteoles elliptic or oblong, 2–2.5 × 1 mm., exceeding
    pedicels or in 1-flowered inflorescences extending to ±
    ⅔ height of the 10–14-grooved calyx; glabrous
    perennial with leaves 1–6.5 × 0.15–0.5 cm., truncate to
    subcordate at base, narrowly acute at apex (U 1)   10. *N. dodecandra*
    Not as above   13

---

* See note at end of *N. erecta* forma *villosa* (p. 24).

13. Leaves larger and thinner, 5–15 × 0.4–1.2 cm., mostly
distinctly auriculate-caudate at base; annual glabrous
or ± scabrid subaquatic herbs . . . .   19. *N. aurita*
Leaves mostly small, linear-oblong to oblong, often stiffer
and more closely placed, 0.3–4.5 × 0.08–1.2 cm. but
often 2 × 0.3 cm. or less; perennial herbs or subshrubs
(one old specimen of *N. parkeri* has linear leaves
up to 6 × 0.4 cm.) 14
14. Leaves, etc., very finely pubescent to densely covered with
short spreading hairs, 0.35–2 × 0.1–0.3 cm.;
inflorescences 1–3-flowered, the peduncles usually
longer than the pedicels; mature flowers campanulate
to subglobose . . . .   15. *N. kilimandscharica*
Leaves, etc., glabrous or rarely minutely puberulous or
ciliolate; inflorescences subsessile to long-peduncled 15
15. Mature flowers with calyx narrower and ± cylindrical, ± 1.5
mm. wide; inflorescences subsessile, the peduncle and
pedicels usually 0.3–3mm. long; leaves 5–12.5 × 1.4–4
mm. . . . .   16. *N. schinzii*
Mature flowers with calyx usually wider, obconic or
campanulate to subglobose, usually over 1.5 mm. wide;
inflorescences usually pedunculate, the peduncles
exceeding the pedicels; leaves larger 16
16. Plants mostly annual; leaves linear to linear-lanceolate,
0.5–4 mm. wide (**K** 4, 7) . . . .   12. *N. parkeri*
Plants perennial; if leaves linear or linear-lanceolate then
rootstock woody or tuberous 17
17. Rootstock consisting of bunches of narrow woody tubers;
leaves 0.4–4.5 × 0.15–1.2 cm. (small, linear, 11 × 0.8 mm.,
sometimes in whorls of 3 in a var.) . . . . . .   14. *N. heptamera*
Rootstock not so far as is known of above nature or not
present in specimen under examination 18
18. Several unbranched or sparsely branched stems 10–40 cm.
tall from a woody rootstock (possibly always tuberous)
mostly in very wet places . . . .   14. *N. heptamera*
Single-stemmed much-branched shrub, 0.45–2.25 m. tall
(or in a var. ± 20-stemmed at base with leaves 0.7–1.5 ×
0.08–0.4 cm.); leaves typically 0.4–3.5 × 0.15–1.2 cm.;
usually in drier places (*stuhlmannii* complex) 19
19. Calyx more ovoid or ellipsoid-cylindric, the ratio of lobes to
tube 0.5: 3; flowers ± persistent; mostly quite shrubby
from a single-stemmed base, 0.45–2.25 m. tall (**T** 7)   13. *N. fruticosa*
Calyx more campanulate or bowl-shaped, the ratio of lobes
to tube 1–1.5: 2.5–3, the lobes larger and triangular;
flowers often deciduous leaving the axes and bracteoles
visible; single-stemmed or many-stemmed (**K** 4, 6, 7;
T 3) 20
20. Coastal plants with ratio of calyx-lobes to tube 1:2.5–3;
single-stemmed shrublets 0.45–1.8 m. tall, much-
branched above; flowers deciduous, plants often with
many persistent flowerless peduncles (**K** 2, **T** 3, 15–75
m.) . . . .   11. *N. stuhlmannii*
Inland plants with ratio of calyx-lobes to tube 1–1.5: 3; at
least some specimens much branched at base; flowers
not deciduous (**K** 4, 6, 1650–1830 m.) . . . . .   15. *N. kilimandscharica*
var. *ngongensis*

FIG. 6. *NESAEA BURTTII*—1, habit, × ²/₃; 2, bracteole, × 4; 3, flower, × 4; 4, corolla, opened out,
× 6²/₃; 5, pistil, × 10; 6, seed, × 24. *N. ASPERA*—7, habit, × 1; 8, bracteole, × 4; 9, flower, ×
4; 10, part of corolla, opened out, × 10; 11, pistil, × 10; 12, fruit, × 4; 13, 14, seed, × 24. *N.
CRASSICAULIS*—15, habit, × ²/₃; 16, flower, one petal removed, × 4; 17, petal, × 6; 18, part of
corolla, opened out, × 10; 19, pistil, × 10; 20, fruit, × 4; 21, seed, × 24. *N. PEDICELLATA*—
22, habit, × ²/₃; 23, flower, × 4; 24, petal, × 6; 25, part of corolla, opened out, ×
10; 26, stamens, × 10; 27, pistil, × 10; 28, fruit (measured dry), × 4. *N. TRIFLORA*—

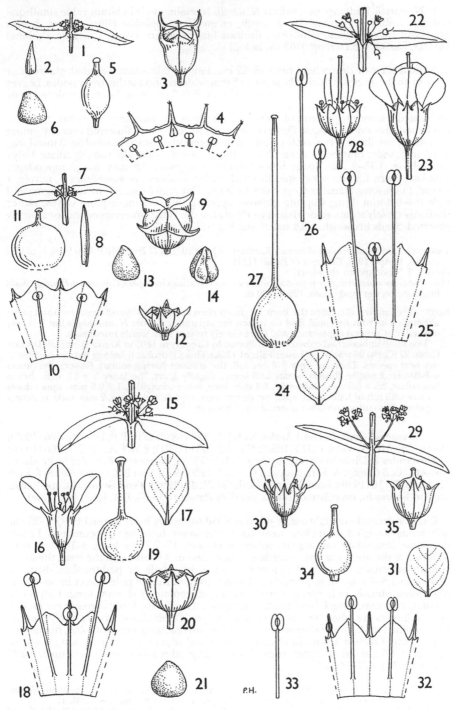

29, habit, × ⅔; 30, flower, × 4; 31, petal, × 6; 32, part of corolla, opened out, × 10; 33, stamen, × 10; 34, pistil, × 10; 35, fruit, × 4. 1–6, from *B.D. Burtt* 3703; 7–11, from *Hooper & Townsend* 2002A; 12–14, from *Hooper & Townsend* 2110; 15–19, from *Drummond & Hemsley* 3500; 20, 21, from *Faulkner* 3416; 22–27, from *Greenway* 9451; 28, from *Revell* 148; 29–35, from *Archbold* 3142. Drawn by Pat Halliday.

1. **N. burttii** *Verdc.*, sp. nov. affinis *N. wardii* Immelman ob habitum valde similisque sed staminibus 4, stigmate haud sessili, etiam *N. sarcophyllae* Hiern affinis, habitu graciliore, inflorescentiis minoribus, floribus 4-meris differt. Typus: Tanzania, Dodoma District, Kazikazi, *B.D. Burtt* 3703 (K, holo.!)

Annual erect or decumbent herb 18–22 cm. long or tall; stems 4-angled, glabrous or scabrid on the angles, ± thick at base, up to 3 mm. wide, rooting at the lower nodes. Leaves decussate, oblong-lanceolate or narrowly elliptic, 0.5–3 cm. long, 3–8 mm. wide, acute or subacute at the apex, ± auriculate-cordate at the base, glabrous, 1-nerved, the nerve impressed above, ± prominent beneath, glabrous or minutely to distinctly hairy, thickened around the edges. Flowers 4-merous in few–many-flowered cymes forming very dense sessile or very shortly pedunculate glomerules; bracts lanceolate, 3 mm. long, 1–1.5 mm. wide; bracteoles linear, 2 mm. long, 0.5 mm. wide, glabrous or ciliate. Calyx subglobose, 1.75–2 mm. long; lobes broadly triangular, 0.8 mm. long; appendages subulate, erect, 1.3 mm. long, apically ciliolate. Petals absent or sometimes 1 , purple, ± round, 1 mm. long. Stamens 4, opposite the lobes. Ovary globose, 2–3-locular; stigma and style 0.4–0.8 mm. long. Capsule globose, splitting into ± 5 incomplete valves, rather distinctly closely striate somewhat as in *Rotala* but the ridges interrupted, not or scarcely exserted. Seeds brownish, ± 0.4 mm. long. Fig. 6/1–6, p. 17.

TANZANIA. Dodoma District: Manyoni, Kazikazi, 10 June 1932, *B.D. Burtt* 3703! & 20.5 km. E. of Itigi station, 11 Apr. 1964, *Greenway & Polhill* 11515!
DISTR. **T** 5; not known elsewhere
HAB. Shallow water fringing seasonal rain-pond and clearings in *Baphia, Combretum, Pseudoprosopis* bushland on wet sandy loam; 1260–1290 m.

NOTE. Brenan first identified the Burtt specimen cited above in 1948 and drew the floral parts; although 4-merous he could find no reason for separating it from *N. sarcophylla* but it is very dissimilar in general appearance and more closely related to *N. wardii* Immelman.
Two small unbranched ephemerals collected by Gregory in 1893 in Kenya (Fort Hall District: Camp 97 (Camp 96 was W. of 'Lower Falls of Thika-Thika')) probably belong here or possibly to the next species. They are only 3–6 cm. tall, the smallest having solitary flowers, the others 4–6-flowered sessile clusters; stems and leaves slightly hairy, the latter 5 × 1.2 mm.; bracts lanceolate, 2.5 × 0.6 mm.; calyx-tube 1.3 mm. long, lobes triangular, 1 × 0.5 mm., appendages narrow with tuft of hairs at the apex, petals not seen: capsule red-brown, 2 mm. wide, punctate-rugulose. It is curious no other material has turned up.

2. **N. aspera** (*Guill. & Perr.*) *Koehne* in E.J. 3: 327 (1882) & in E.P. IV. 216: 226 (1903); Hiern, Cat. Afr. Pl. Welw. 1: 375 (1898); V.E. 3(2): 651 (1921); F.W.T.A., ed. 2, 1: 166 (1954); A. Fernandes & Diniz in Garcia de Orta 6: 105 (1958); Pohnert & Roessler in Prodr. Fl. SW.-Afr. 95: 5 (1966); A. Fernandes in C.F.A. 4: 183 (1970) & in F.Z. 4: 289 (1978) & in Fl Moçamb. 73: 16 (1980); Immelman in Bothalia 21: 42 (1991). Type: Senegal, Cape Verde, around Kounoun, no collector stated, possibly *Perrottet* 338 (G, BM, iso.?) (see note)

Erect or decumbent glabrous or often scabrid to shortly hairy annual herb, 4–25 cm. tall; stems 4-angled, branched near base, the lower branches procumbent. Leaves decussate, lanceolate, oblong-lanceolate or elliptic, 0.7–2.7 cm. long, 3–10 mm. wide, subacute at the apex, the upper cordate, subcordate or ± rounded at the base, the lower cuneate-attenuate. Flowers 4(–6)-merous, 3–few in sessile or pedunculate dichasia; peduncles 1.5–4 mm. long; pedicels 1–1.5 mm. long or that of pole flower up to 4 mm.; bracteoles whitish with green nerve, oblong-lanceolate, 2–4 mm. long. Calyx-tube cupular, ± 1.5 mm. long; lobes broadly triangular, 0.3–0.5 mm. long; appendages erect, ± twice as long as the lobes, sometimes ciliolate at apex. Petals absent or very rarely present. Stamens inserted below middle of the calyx-tube and reaching apices of lobes, filaments ± 1 mm. long. Ovary ± 1 mm. in diameter; style and stigma 0.3–0.6 mm. long. Capsule globose, ± 1.5 mm. in diameter, scarcely exceeding calyx-lobes. Seeds brownish, ± 0.25 mm. long. Fig. 6/7–14, p. 17.

TANZANIA. Tabora District: S. of Pozo Moyo, 8 km. from Kaliua [Kaliuwa], 22 June 1980, *Hooper, Townsend & Mwasumbi* 2110! & 16 June 1960, *Hooper, Townsend & Mwasumbi* 2002A!
DISTR. **T** 4; Senegal, Mozambique, Zimbabwe, Botswana, Angola, South Africa (N. Natal) and Namibia
HAB. Damp depressions and abandoned paddy fields; ± 1300 m.

SYN. *Ammannia aspera* Guill. & Perr., Fl. Seneg. Tent.. 304 (1833); Hiern in F.T.A. 2: 480 (1871); F.W.T.A. 1: 144 (1927)

NOTE. The specimens cited, one wrongly named as *Ammannia prieuriana*, the other as *A. senegalensis*, are the only ones seen from the Flora area and the species has not been previously recorded; it is a glabrous form. Immelman states that no *Leprieur* or *Perrottet* specimen can be found in Paris. *Perrottet* 338 is at the BM (with minute fragments at K), but only part of it is *N. aspera*, and also at G; the BM sheet is labelled Walo so is presumably not strictly type material; *Perrottet* 403 (P) labelled Casamance is strictly not type material since that locality is not cited and moreover the specimen is a slender form of *N. crassicaulis* (Guill. & Perr.) Koehne.

3. **N. crassicaulis** (*Guill. & Perr.*) *Koehne* in E.J. 3: 324 (1882); E.P. IV. 216: 225, fig. 43 A (1903); R.E. Fries, Wiss. Ergebn. Schwed. Rhod. Kongo Exped. 1: 165 (1914); Koehne in V.E. 3(2): 651 (1921); H. Perrier in Fl. Madag. 147, Lythracées: 7 (1954); F.W.T.A., ed. 2, 1: 166 (1954) & 760 (1958); A. Fernandes & Diniz in Garcia de Orta 6: 104 (1958); Pohnert & Roessler in Prodr. Fl. SW.-Afr. 95, Lythraceae: 6 (1966); Boutique in F.C.B., Lythraceae: 18 (1967); A. Fernandes in C.F.A. 4: 182 (1970) & in F.Z. 4: 287 (1978) & in Fl. Moçamb. 73: 14 (1980). Types: Senegal, Cape Verde, Khann & Cayor, Ghielcouil Springs, *Perrottet* 337 (P, syn.)

Annual or perennial completely glabrous prostrate or ascending subsucculent or aquatic herb, 0.07–0.5(–2) m. long with ± thick reddish brown ± 4-angled stems rooting at the lower nodes. Leaves obovate-oblong to oblanceolate or lanceolate-elliptic, 0.5–6 cm. long, 0.3–2 cm. wide, obtuse or ± rounded at the apex, cuneate at the base, 1-nerved, sessile. Cymes lax, 1–7(–many)-flowered; peduncle 0–2(–15) mm. long; bracteoles linear-lanceolate, 0.5–1 mm. long; pedicels 2–6 mm. long. Flowers 4-merous. Calyx ± globose, 2–2.5 mm. long; lobes triangular, 0.5–0.6 mm. long; appendages equalling the lobes. Petals blue or mauve, 4 or sometimes wanting, broadly elliptic to round, 1–1.5 mm. long. Stamens 4–6(–7), included. Ovary globose, ± 1 mm. diameter, 2-locular; style 0.5–1 mm. long. Capsule subglobose, ± 2 mm. in diameter, equalling the calyx-lobes. Seeds brownish, concave-convex, ± 0.5 mm. long. Fig. 6/15–21, p. 17.

UGANDA. Busoga District: SE. of Buluba leper colony, 3 Oct. 1952, *G.H.S. Wood* 469!
KENYA. Kisumu, 2 Feb. 1960, *McMahon* K302!
TANZANIA. Tanga District: Ngomeni, 28 July 1953, *Drummond & Hemsley* 3500!; Kigoma District: Malagarasi R., 18 Sept. 1952, *Lowe*! & Uvinza, 18 Sept. 1952, *Lowe* in *E.A.H.* 11048!; Zanzibar I., Mwera swamp, 19 Aug. 1960, *Faulkner* 2485! & 20 Aug. 1964, *Faulkner* 3416!
DISTR. U 3; K 5; T 3, 4; Z; Senegal, Mali, Guinea Bissau, Zaire, Central African Republic, Zambia, Mozambique, Zimbabwe, Botswana, Caprivi Strip and Angola (see note)
HAB. Swampy places; 0–1125 m.

SYN. *Ammannia crassicaulis* Guill. & Perr., Fl. Seneg. Tent. 1: 303 (1833); Hiern in F.T.A. 2: 479 (1871)
[*A. sarcophylla* sensu U.K.W.F.: 154 (1974), *non* Hiern]
NOTE. Sterile *Ludwigia stolonifera* (Guill. & Perr.) Raven has been confused with this but the petioles are longer and more distinct and the leaf-nervation different. The reported occurrence in Madagascar is, as explained by Perrier de la Bâthie, due to a confusion; the Bojer specimen concerned has no indication of locality and almost certainly came from Zanzibar.

4. **N. pedicellata** *Hiern* in F.T.A. 2: 472 (1871); Koehne in E.J. 3: 329 (1882) & 6: 405 (1885), in P.O.A. C: 285 (1895), in E.P. IV. 216: 229, fig. 45A (1903) & in V.E. 3 (2): 651, fig. 286A (1921); A. Fernandes in Bol. Soc. Brot., sér. 2, 52: 4(1978). Types: Zanzibar*, probably the island, *Kirk* s.n. & *Kirk* 6 (K, syn.!)

Erect or ascending glabrous herb 1–50 cm. tall or long, the short specimens erect; stems thickish, ± fleshy, 4-angled, sparsely branched, often rooting at basal nodes. Leaves decussate, narrowly elliptic or obovate, 0.6–4.2 cm. long, 0.2–1.5 cm. wide, ± obtuse at the apex, acute to obtuse at the base, ± 1-nerved. Flowers 4–numerous, heterostylous, probably trimorphic in (1–)3–5(–9)-flowered dichasia; lower peduncles 0.1–1.2 cm. long, others 0.5–2.5 mm.; pedicels 2.5–6 mm. long; bracteoles subulate or lanceolate, scarcely 1 mm. long. Calyx ± 2.5 mm. long; lobes ± 0.8 mm. long, the alternating appendages 0.5 mm. long, spreading. Petals mauve, wine-red, or blue (*fide* Greenway), obovate, 3–4 mm. long, 1.5–4 mm. wide. Stamens 8. Long-styled flowers with longer stamens ⅔-exserted, shorter reaching tips of calyx-lobes; style ± 4–5 mm. long, overtopping the longer sepals. Intermediate-styled flowers with longer stamens just overtopping the stigma, shorter reaching just below stigma; style ± 2.5 mm. long. Short-styled flowers with longer stamens 4.5 mm. long, shorter 3.5 mm. long; style ± 1.25 mm. long. Capsule globose, not or scarcely exserted from the calyx. Fig. 6/22–28, p. 17.

* Neither specimen specifies the actual island; one is labelled Prov. Zanguebar and the other Flora Zanguebarica and could have been from mainland coast.

KENYA. Kwale District: Gongoni Forest, 5 Jan. 1992, *Luke* 3043!; Lamu District: Lake Mukunguya, 5
Nov. 1957, *Greenway & Rawlins* 9451!
TANZANIA. Near Tanga, 8 Aug. 1932, *Geilinger* 1230!; Uzaramo District: 12 km. ESE. of Dar es Salaam,
between Mjimwema–Mboamaji, Mwera swamp, 8 Oct. 1975, *Wingfield* 3182! & Dar es Salaam, May
1952, *Revell* 148!; Zanzibar I., Josani Forest, July 1972, *Robins* 26!
DISTR. K 7; T 3, 6; Z; not known elsewhere
HAB. Seasonally flooded freshwater swamps and swamps next to permanent lakes and streams with
grasses, Cyperaceae, *Aeschynomene*, *Ipomoea aquatica*, etc.; forest on loam on coral and drainage
ditches; persisting as a weed in rice-fields; 0–150 m.

SYN. *N. procumbens* A. Peter in Abh. Ges. Wiss. Göttingen N.F. 13(2): 87 (1928). Type: none given but
Peter 44893, 52047 & 52048 cited by A. Fernandes in Bol. Soc. Bot. Brot., sér. 2, 52: 4 (1978) can
probably be taken as syntypes.

NOTE. Greenway in notes to 9451 cited above suggests it is perennial but I think this unlikely
although individuals may persist more than a year in swamps which do not dry up; he also claims
the flowers are blue. There is no doubt that this is very close to *N. crassicaulis* but it is not difficult to
distinguish and, moreover, has a restricted eastern coastal distribution unlike the widespread *N.
crassicaulis*. The ovary anatomy is that of *Ammannia* but the two genera need further study
worldwide.

5. **N. triflora** (*L.f.*) *Kunth*, Nov. Gen. Sp. 6: 191 (1824); Wight, Ic. Pl. Ind. Or., t. 2559
(1840); Koehne in E.J. 3: 330 (1882) & in E.P. IV. 216: 230 (1903); H. Perrier in Fl. Madag.
147, Lythracées: 8 (1954); Coode in Fl. Masc. 95, Lythracées: 4, t. 2/5–8 (1990). Type:
supposedly from America, no collector cited but probably a *Commerson* specimen from
Mauritius* (P, holo., G, iso. (microfiche!))

Annual or short-lived perennial herb or aquatic (*fide* Perrier), with glabrous decumbent
stems 15–70 cm. long (over 1 m. in subsp. *lupembensis*), 4-angled, often rooting at the
nodes. Leaves often reddish, narrowly oblong to oblong-ovate or oblong-lanceolate,
1–3.5(–5.5) cm. long, 0.4–1.4(1.8) cm. wide, obtuse to ± acute at the apex, rounded, truncate
or rarely subcordate at the base, glabrous, ± uninerved; petiole up to 1.5 mm. long.
Flowers 4–5(–6)-merous in axillary pedunculate 3(–5)-flowered cymes (11–17-flowered in
subsp. *lupembensis*); peduncles 1–2.1(–2.5–?4**)cm. long, glabrous or puberulous;
bracteoles lanceolate, linear-subulate or ± elliptic, 2–5 mm. long, ± folded; pedicels ± 1
mm. long. Calyx-tube greenish purple, obconic, becoming subglobose, 2.5–3 mm. long.
8–10-nerved; lobes broadly triangular, ± 1 mm. long, acute, appendages, 1 mm. long,
3-denticulate at apex. Petals rose, lilac or rose-purple, obovate, 5–6 mm. long, 3–3.5 mm.
wide, rounded at apex, narrowed into a short claw at the base. Stamens(4), 8, 10 (or 12),
3–5 mm. long. Ovary globose, ± 1 mm. diameter; style purple, ± 5 mm. long in long-styled
flowers, ± 1–2 mm. in short-styled flowers. Capsule globose, ± 5 mm. in diameter.

subsp. **triflora**

Stems 15–70 cm. long. Leaves 1–3.5 cm. long, 0.4–1.4 cm. wide, rounded to narrowed at the base.
Cymes 3(–5)-flowered. Fig. 6/29–35, p. 17.

KENYA. Lamu District: 3 km. inland from Kiunga on road to Mararani, Badar water pan, 5 Apr. 1980,
Gilbert & Kuchar 5881!
TANZANIA. Tanga District: 8 km. on Tanga–Pangani road, 21 Apr. 1973, *Faulkner* 4769C! & "5 m"
[? miles or m. altitude] Tanga, 3 Sept. 1987, *Archbold* 3142!
DISTR. K 7; T 3; Comoro Is., Madagascar, Mascarene Is. and Sri Lanka (*fide* Koehne)
HAB. With sedges, *Hydrolea*, etc., on black soil, edges permanent water-hole with *Pistia*, *Neptunia*,
sedges, *Dactyloctenium geminatum*, etc.; rice-fields; ± sea-level to perhaps ± 50 m.

SYN. *Lythrum triflorum* L.f., Suppl. Pl.: 249 (1782)

NOTE. These appear to be the first records of this species from the African continent — apart from
puberulous peduncles it seems ± indistinguishable from material elsewhere. Coode gives the style
length as ± 5 mm. and Perrier as 1 mm.; Koehne also states 'twice as long as ovary'. The species is
clearly heterostylous and both states are actually represented in the Mascarene material which
consists of more specimens than from any other area. *Faulkner* 4769 was a mixed gathering of the
above taxon together with *N. pedicellata* Hiern, *Hydrolea sansibarica* Gilg and another *Nesaea*
(4769B) which is probably only a variant of *N. triflora* although at first I thought it might represent a
distinct species. It has mostly smaller leaves and more divaricate inflorescences with smaller bracts
and flowers. The dimensions given below will modify the above description. Stems 13 cm. tall from
decumbent base; leaves sessile, narrowly elliptic-oblong, 0.9–2.2 cm. long, 2.5–5 mm. wide; flowers

---

* See DC., Revue Fam. Lythraires: 10, 11 (1826).
** Koehne mentions up to 40 mm. in his key.

4-merous in 3–7-flowered small ± lax dichasia; peduncles 2.5–5 mm. long; pedicels 0.5–1(–1.5) mm. long; bracteoles lanceolate, 1–2 mm. long; calyx-tube 1–2 mm. long, 8-ribbed with lobes 0.5 mm. long; style 0.6 mm. long. Only more material will help decide the status of this.

subsp. **lupembensis** *Verdc.*, subsp. nov. a subsp. *triflora* caule elongato, foliis majoribus, cymis plurifloris differt. Typus: Tanzania, Ulanga District, Ifinga, *Schlieben* 1326 (BM, holo.!, BR, iso.!)

Well over 1 m. long, the lower decumbent part of main stem with long roots at most nodes and ± terete, much-branched with many presumably ± erect branches. Leaves up to 5.5 cm. long, 1.8 cm. wide, mostly truncate or subcordate at the base. Cymes up to 11–17- or more-flowered.

TANZANIA. Ulanga District: Ifinga, 10 Oct. 1931, *Schlieben* 1326!
DISTR. **T** 6; not known elsewhere
HAB. River bank; 700 m.

NOTE. The sheet had been sent out from Berlin determined *Nesaea triflora* (L.f.) Kunth. var. vel. nov. spec.; it certainly appears very different from typical *N. triflora* but is clearly closely related. The two specimens seen have styles 5 mm. long.

6. **N. radicans** *Guill. & Perr.*, Fl. Seneg. Tent. 1: 306, t. 70 (1833); Hiern in F.T.A. 2: 474 (1871); Koehne in E.J. 3: 330 (1882); Gilg in P.O.A. C: 285 (1895); Grandidier, Atlas Hist. Pl. Madag., t. 360 (1896); Hiern in Cat. Afr. Pl. Welw. 1: 376 (1898); Koehne in E.P. IV. 216: 232 (1903); Engl., V.E. 3(2): 651 (1921); F.W.T.A., ed. 2, 1: 166 (1954); H. Perrier, Fl. Madag. 147, Lythracées: 7 (1954); A. Fernandes & Diniz in Garcia de Orta 4: 394 (1956) & 6: 107 (1958); Boutique, F.C.B., Lythraceae: 17 (1967); A. Fernandes in C.F.A. 4: 188 (1970) & in F.Z. 4: 298 (1978) & in Fl. Moçamb. 73: 26 (1980); Immelman in Bothalia 21: 43 (1991). Type: Senegal, Cayor, *Perrottet* 341 (P, lecto., BM, isolecto.!)

Erect spreading or procumbent perennial herb (0.08–)0.2–1.2 m. long; stems glabrous, greyish-pubescent or glabrescent occasionally becoming woody at base, often rooting at the nodes particularly in wet habitats. Leaves decussate, oblong-elliptic to oblong-lanceolate, elliptic-ovate or rarely ovate, 1–5.5 cm. long, 0.4–2.5 cm. wide, rounded to subacuminate at the apex, rounded to narrowed at base, cartilaginous and sometimes ciliate at the purplish margins, uninerved or sometimes with distinct lateral nerves, glabrous to pubescent, sessile or petiole 1–3 mm. long. Flowers 5–6(–7)-merous in many-flowered dichasia; peduncles 0.7–5.5 cm. long; pedicels 0.5–3 mm. long; bracteoles green flushed red, or whitish, 2–4, leafy, 3–10 mm. long and wide, shortly acuminate, cordate, venose, glabrous or ciliate; inner bracteoles linear, 3–4 mm. long. Calyx campanulate, 3–4 mm. long, 10–12(–14)-ribbed; lobes reddish, triangular, ± 0.5 mm. long; appendages 1–5 mm. long, erect, ciliate at the apex. Petals pink, rose-red, mauve or purplish-magenta, obovate, 2.5–5.5 mm. long, up to 3 mm. wide. Stamens pink, usually twice as many as the calyx-lobes, ± unequal, exserted. Ovary subglobose, 1–1.5 mm. in diameter, subglobose; style 6–8 mm. long. Seeds yellowish, concave-convex, 0.4 mm. long.

var. **radicans**
Plant quite or practically glabrous. Fig. 7.

UGANDA. W. Nile District: Koboko, Mar. 1938, *Hazel* 436!
KENYA. Machakos District: Kibwezi, 28 Nov. 1910, *Scheffler* 380!; Kwale District: near Mrima Hill, 1 Feb. 1983, *S.A. Robertson* 3512! & Mombasa–Shimoni, *Whyte*!
TANZANIA. Shinyanga–Tinde, 22 May 1951, *Backlund* 4!; Lushoto, 25 Oct. 1959, *Semsei* 2936!; Kigoma District: 4.8 km. on Kigoma–Kasulu road, 11 July 1960, *Verdcourt* 2792!; Pemba I., Mkoani, 10 Aug. 1929, *Vaughan* 483!
DISTR. **U** 1; **K** 4, 7; **T** 1–8; **Z**; **P**; widespread in tropical Africa and Madagascar
HAB. Swampy ground ("mbugas"), grassy places on lava, pool edges, trickles in sandy gullies, seasonal sand-rivers, also a weed in rice-fields; 0–1500 m.

var. **floribunda** (*Sond.*) *A. Fernandes* in Bol. Soc. Brot., sér. 2, 48: 117 (1974) & in F.Z. 4: 299 (1978) & in Fl. Moçamb. 73: 27 (1980); Immelman in Bothalia 21: 43 (1991). Type: South Africa, near Durban [Port Natal], Omblas R., *Drège* (TCD, holo., PRE, S, iso.!)

Plant ± sparsely to densely pubescent, particularly upper parts of stems.

UGANDA. Busoga District: Bugweri, 1.6 km. W. of Busesa, Nabukolyo A.L.G. Plantation, 1 Aug. 1952, *G.H.S. Wood* 305!; Mbale District: Budadiri, 16 Jan. 1931, *Hill* 8!; Mengo District: km. 85 on Kampala–Masinde road, 20 Aug. 1960, *Lind* 2731!
KENYA. Northern Frontier Province: Moyale, 18 Apr. 1952, *Gillett* 12844!; Kiambu, 10 Aug. 1930, *Napier* 385!; Kilifi District: Mida, Arabuko-Sokoke Forest Reserve, 3 Dec. 1961, *Polhill & Paulo* 891!
TANZANIA. Musoma District: Bologonja R., 19 Aug. 1962, *Greenway* 10765!; Tanga District: 24 km. SW. of Kwale, 27 Aug. 1953, *Drummond & Hemsley* 4021!; Singida Lake, 27 Apr. 1962, *Polhill & Paulo* 2207!; Zanzibar I., Mkokotoni, Mwera swamp, 27 June 1960 & 15 July 1960, *Faulkner* 2618!

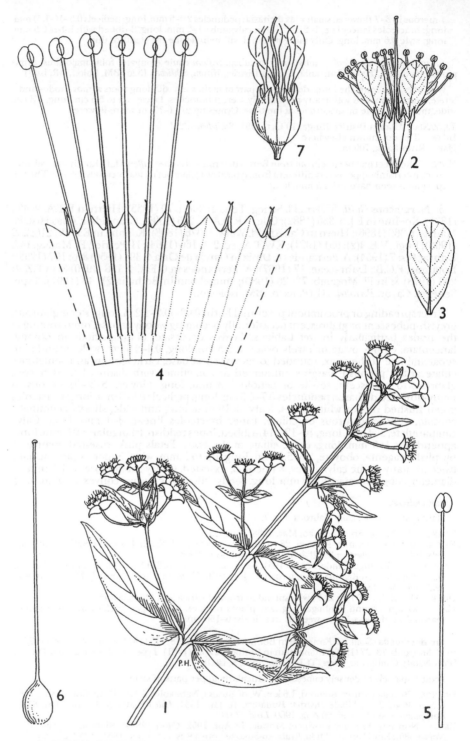

FIG. 7. *NESAEA RADICANS* var. *RADICANS*—1, habit, × ²⁄₃; 2, flower, one petal removed, ×
4; 3, petal, × 6; 4, corolla, opened out, × 10; 5, stamen, × 8; 0, pistil, × 8; 7, fruit, × 4 All from
B.D. Burtt 3805. Drawn by Pat Halliday.

DISTR. U 3, 4; K 1, 3, 4, 6, 7; T 1–5, 7; Z; widespread in tropical (mainly eastern) and southern Africa
HAB. Marshes, swampy grassland, muddy ditches, riverbanks, reed-beds, sand rivers, rice-fields, etc.; 0–1750 m.

SYN. *Nesaea floribunda* Sond. in Fl. Cap. 2: 517 (1862); Hiern in F.T.A. 2: 474 (1871); Koehne in E.J. 3: 331 (1882); Hiern, Cat. Afr. Pl. Welw. 1: 376 (1898); Koehne in E.P. IV. 216: 231 (1903); Thonner, Blutenpfl. Afr., t. 111 (1908) & ed. 2, Fl. Pl. Afr., t. 110 (1915); V.E. 3(2): 651 (1921); A. Fernandes in C.F.A. 4: 189 (1970)

7. **N. erecta** *Guill. & Perr.*, Fl. Seneg. Tent. 1: 305, t. 69 (1833); Hiern in F.T.A. 2: 474 (1871); Koehne in E.J. 3: 331 (1882); Gilg & Koehne in P.O.A. C: 285 (1895); Hiern, Cat. Afr. Pl. Welw. 1: 371 (1898); Koehne in E.P. IV. 216: 231 (1903); Engl., V.E. 3(2): 651 (1921); A. Peter in Abh. Ges. Wiss. Göttingen N.F. 13(2): 87 (1928); F.P.S. 1: 141, fig. 86 (1950); F.W.T.A., ed. 2, 1: 166 (1954) & 760 (1958); A. Fernandes & Diniz in Garcia de Orta 4: 396 (1956) & 6: 107 (1958); Heriz-Smith, Wild Fl. Nairobi Nat. Park: 50 (1962); Boutique in F.C.B., Lythraceae: 17 (1967); A. Fernandes in C.F.A. 4: 188 (1970); U.K.W.F.: 154 (1974); A. Fernandes in F.Z. 4: 299 (1978); M.G. Gilbert in Fl. Eth., ined. Types: Senegal, Cayor, near Laybar, *Perrottet* 343 (P, syn.) & bank of R. Casamance near Samatite, *Perrottet* 342 (P, syn.)

Erect mostly glabrous or sometimes very shortly pubescent or shortly hispidulous annual herb, usually ephemeral, 2.5–15(–35) cm. tall with branched or unbranched 4-angled often crimson stems. Leaves decussate, sessile, linear to narrowly oblong, lanceolate or oblanceolate, 0.8–3.5 cm. long, 1.5–8(–10) mm. wide, acute to obtuse at apex, cuneate at the base, uninerved. Flowers 4–6-merous in solitary axillary (3–)5–7(–many)-flowered head-shaped dichasia, surrounded at the base by 2 greenish, broadly cordate acuminate nervose involucral bracts (3.5–)7.5–8 mm. long, (3–)6–7.5 mm. wide; peduncles 0.3–1.5 cm. long, glabrous or with patent hairs; bracteoles linear-lanceolate, 3–4 mm. long, 1 mm. wide. Calyx campanulate, 1.2–2 or reported to attain 2.5–3.5 mm. long, 1.2–1.8 mm. wide; lobes triangular, 0.5 mm. long, the alternating appendages erect, linear, 1–2.5 mm. long, ± bent inwards, glabrous or ciliate. Petals rose, lilac, magenta or violet, obovate, 2.5–3 mm. long, clawed. Stamens unequal, as many or twice as many as the calyx-lobes, up to 2 mm. long, exserted. Ovary subglobose, 1–2 mm. in diameter, 2–4-locular; style exserted 2.5–3.5 mm., overtopping the stamens. Capsule globose, 2.5 mm. diameter, enclosed in the calyx. Seeds brownish, ± 0.3 mm. long.

forma **erecta**
Plant glabrous or ± hispidulous with very short hairs. Fig. 8/1–11, p. 25.

UGANDA. Nile Province, prob. Nov. 1905, *Dawe* 919!*
KENYA. Northern Frontier Province: Moyale, 10 July 1952, *Gillett* 13564!; Laikipia District: near Rumuruti, June 1975, *Powys* 45!; Nairobi National Park, Impala Point, 21 Jan. 1962, *Verdcourt* 3242!
TANZANIA. Tabora District: Kaliua [Kaliuwa], 16 June 1980, *Hooper et al.* 2002B!; Kilosa District: 21 km. on Mvumi–Mandege track, 30 May 1978, *Thulin & Mhoro* 2698!; Morogoro District: 12.8 km. NE. of Kingolwira station, 19 Aug. 1955, *Welch* 317!; Uzaramo District: Msasani, July 1939, *J. Vaughan* 2838!
DISTR. U 1; K 1, 3, 4; T 2, 4, 6; Senegal, Gambia, Mali, Nigeria, Zaire, Sudan, Ethiopia, Zambia, Mozambique, Zimbabwe, Angola and Madagascar
HAB. Wet flushes in edaphic grassland and seepage zones, thin red soil on lava outcrops and granite, vleis and rivulets on damp mud, usually associated with small Cyperaceae and other ephemerals, also in abandoned rice cultivations; 75–1800 m.

SYN. *N. humilis* Klotzsch in Peters, Reise Mossamb., Bot. 1: 68 (1861). Type: Mozambique, Rios de Sena, *Peters* (B, holo.,†, K, iso.!)
N. racemosa Klotzsch in Peters, Reise Mossamb., Bot. 1: 68 (1862). Type: Madagascar, no locality or collector cited (note: H. Perrier, Fl. Madag. 147, Lythracées: 9 (1954) states that *N. erecta* has not been refound)
N. erecta Guill. & Perr. forma *glabra* Koehne in E.J. 3: 332 (1882) & in E.P. IV. 216: 232 (1903). Types: Ethiopia, Schire, *Quartin Dillon & Petit* (P, syn.) & Sudan, Bongo, Gir, *Schweinfurth* 2530 (B, syn.†) & Mozambique, Rios de Sena, *Peters* (B, syn.†, K, isosyn.!)
N. erecta Guill. & Perr. forma *hirtella* Koehne in E.J. 3: 232 (1882) & in E.P. IV. 216: 232 (1903), *nom. illegit.* Type: as for *N. erecta*

forma **villosa** *A. Fernandes* in Bol. Soc. Brot., sér. 2, 52: 5 (1978); Vollesen in Opera Bot. 59: 52 (1980). Type: Tanzania, Kilwa District, Selous Game Reserve, ± 3 km. NNW. of Kingupira, *Vollesen* in M.R.C. 2570 (DSM, holo., C, EA!, K! WAG, iso.)

* Mentioned in Dawe, 'Rep. Bot. Miss. Uganda Prot.: 46 (1906)' as a weed in moist swampy places.

Plant fairly to quite densely pubescent with distinct hairs.

TANZANIA. Tanga District: Sawa, 21 Aug. 1958, *Faulkner* 2179!; Pangani District: Mwera, Stahaabu, 29 Aug. 1957, *Tanner* 3666!; Uzaramo District: Dar es Salaam, Manzese, 26 July 1968, *Harris & Mwasumbi* 2058!; Kilwa District: 3 km. NNW. of Kingupira, 25 Aug. 1976, *Vollesen* in M.R.C. 3971!
DISTR. **T** 3, 6, 8; not known elsewhere
HAB. Seepages in wooded grassland, black cotton soil, salt-marshes, rice-fields; 0–125 m.

NOTE. *Semsei* 2930 (Lushoto District, Korogwe, Lwengera [Luengera] valley, 27 Sept. 1959) seems to be a small form of *Nesaea radicans* Guill. & Perr. var. *floribunda* (Sond.) A. Fernandes; further work may show forma *villosa* is an ephemeral form of this.

8. **N. cordata** *Hiern* in F.T.A. 2: 475 (1871); Oliv. in Trans. Linn. Soc. 29: 74, t. 40B (1873); Koehne in E.J. 3: 332 (1882); Gilg in P.O.A. C: 285 (1895); Hiern, Cat. Afr. Pl. Welw. 1: 376 (1898); Koehne in E.P. IV. 216: 232 (1903); V.E. 3(2): 651 (1921); F.P.S.: 141 (1950); F.W.T.A., ed. 2, 1: 166 (1954); A. Fernandes & Diniz in Garcia de Orta 4: 392 (1956) & 6: 106 (1958); Pohnert & Roessler in Prodr. Fl. SW.-Afr. 95, Lythraceae: 6 (1966); Boutique, F.C.B., Lythraceae: 15 (1967); A. Fernandes in C.F.A. 4: 187 (1970) & in F.Z. 4: 302 (1978) & in Bol. Soc. Brot., sér. 2, 52: 5 (1978) & in Fl. Moçamb. 73: 34 (1980); Immelman in Bothalia 21: 43 (1991). Type: Uganda, W. Nile District, Madi Swamp, *Grant* (K, lecto.!)

Erect annual or ephemeral ± branched herb 3–30 cm. tall; stems subglabrous to ± hairy, the upper ones 4-angled or narrowly 4-winged. Leaves sessile, decussate, ovate-lanceolate or sometimes lanceolate, 0.5–2.5 cm. long, 0.5–1 cm. wide, acute at the apex, truncate to cordate at the base (at least in upper leaves), glabrous, ± asperulous or ± hairy, 1-nerved. Flowers 4(–6)-merous, subsessile, in ± numerous ± 5-flowered ± capitate dichasia surrounded at the base by 2 imbricate broadly cordate acuminate cymbiform bracts 4–7 mm. long and wide, often ± hairy outside, sometimes white, veined pink beneath; bracteoles linear to elliptic, 1.5–3.3 mm. long, acute, ciliate; peduncles 0.5–2.7 cm. long. Calyx-tube campanulate, 1.75–2 mm. long, 8- or 12-ribbed; lobes often red, triangular, 0.5 mm. long, the intermediary appendages erect, ± 1 mm. long, ciliate at the apex. Petals rose, mauve or pink, obovate, 1.5–2 mm. long, deciduous, ? occasionally lacking. Stamens 4–6, rarely 8–12, exserted. Ovary subglobose, ± 1 mm. in diameter, 2(–3)-locular; style 1.5–2 mm. long. Capsule subglobose, 1.5–1.7 mm. long, included. Fig. 8/12–21.

UGANDA. W. Nile District: Madi swamp, 14 Dec. 1862, *Grant*!; Teso District: Atira, Dec. 1925, *Maitland* 1317!
TANZANIA. Dodoma District: Manyoni, Kazikazi, 18 May 1932, *B.D. Burtt* 3687!; Iringa District: Ruaha National Park, Msembe [Msembi]–Mbage [Mbagi] track near junction of Causeway Track, 18 Apr. 1970, *Greenway & Kanuri* 14379!; Njombe District: Iyayi, 15 Apr. 1962, *Polhill & Paulo* 2009!; Tunduru District: 1 km. E. of Songea District boundary, 6 June 1956, *Milne-Redhead & Taylor* 10593!
DISTR. **U** 1, 3, ?4; **T** 2, 4–8; Central African Republic, Mali, Ivory Coast, Ghana, Nigeria, Sudan, Mozambique, Malawi, Zambia, Zimbabwe, Angola, Botswana, Namibia and South Africa
HAB. Areas of impeded drainage, boggy grassland, damp sandy places, with grasses and sedges in seepage zones, in *Combretum, Commiphora* open bushland, *Combretum, Xeroderris, Adansonia, Excoecaria* bushland, *Sterculia, Isoberlinia, Combretum, Terminalia, Acacia* woodland, marshy places in *Brachystegia* woodland; 710–1620 m.

SYN. [*N. erecta* sensu Thomson in Speke, Journ. Disc. Source Nile, App.: 634 (1863), *non* Guill. & Perr.]
N. cordata Hiern forma *villosa* Koehne in E.J. 29: 166 (1900) & in E.P. IV. 216: 232 (1903). Type: Malawi, without locality, *Buchanan* 423 (B, holo.†, E, iso.!)
N. sagittata A. Peter in Abh. Ges. Wiss. Göttingen N.F. 13(2): 86, fig. 17 (1928). Type: Tanzania, Buha District, Mkuti R., Musoso [Msosi], *Peter* 37208 (B, holo.)

NOTE. No one appears to have suggested that this should be combined with *N. erecta* but the two are certainly extremely close and some specimens are intermediate, e.g. *Dummer* 2908 (Uganda, Mengo District, Jumba, Jan. 1916) has narrow leaves but they are ± truncate at base; over large areas of southern Africa *N. cordata* appears quite separable and I have left it so but investigation is certainly needed and experimental cultivation would be easy.

9. **N. linearis** *Hiern* in F.T.A. 2: 475 (1871); Koehne in E.J. 3: 333 (1882) & in E.P. IV. 216: 233, fig. 45C (1903) & V.E. 3(2): 651, fig. 286C (1921); Garcia in Est. Ens. Doc. Junta Invest.

FIG. 8. *NESAEA ERECTA* forma *ERECTA*—**1**, habit, × ²⁄₅; **2**, bracteole, × 4; **3**, flower, one petal removed, × 4; **4**, petal, × 6; **5**, part of corolla, opened out, × 10; **6, 7**, stamens, × 10; **8**, pistil, × 10; **9**, fruit, × 4; **10, 11**, seed, × 24. *N. CORDATA*—**12**, habit, × ²⁄₅; **13**, bracteole, × 4; **14**, flower, one petal removed, × 4; **15**, petal, × 6; **16**, part of corolla, opened out, × 10; **17**, stamen, × 10; **18**, pistil, × 10; **19**, fruit, × 4; **20, 21**, seed (? immature), × 36. *N. LINEARIS*—**22**, habit, × ²⁄₅; **23**, bracteole, × 4; **24**, flower, × 4; **25**, petal, × 6; **26**, part of

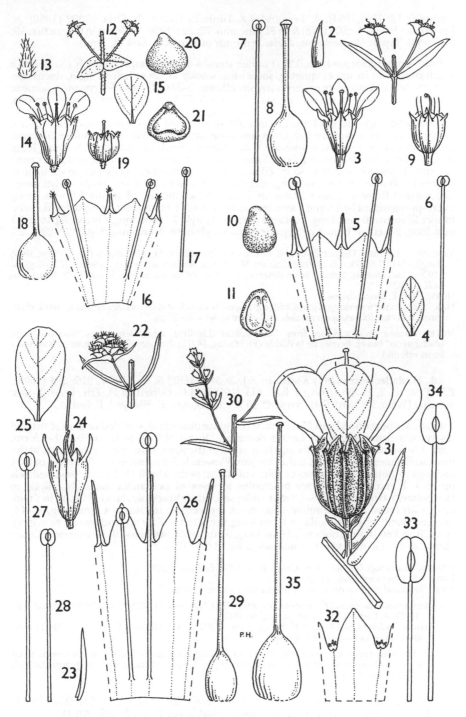

corolla, opened out, × 10; **27, 28**, stamens, × 10; **29**, pistil, × 10. *N. DODECANDRA*—**30**, habit, × ⅔; **31**, flower, × 4; **32**, part of corolla, opened out, × 10; **33, 34**, stamens, × 10; **35**, pistil, × 8. 1–11, from *Thulin & Mhoro* 2698; 12–18, from *Milne-Redhead & Taylor* 9938; 19–21, from *Greenway & Kanuri* 14497; 22–24, 26–29, from *Wingfield* 4167; 25, from *Wingfield* 4192; 30–35, from *Liebenberg* 208. Drawn by Pat Halliday.

Ultramar 12: 159 (1954); A. Fernandes & Diniz in Garcia de Orta 4: 392 (1956); A. Fernandes in F.Z. 4: 297 (1978) & in Fl. Moçamb. 73: 22 (1980); Vollesen in Opera Bot. 59: 52 (1980). Types: Mozambique, *Forbes* & mouth of Zambezi, *Kirk* (K, syn.!)

Erect annual glabrous herb, 0.25–1 m. tall; stems 4-angled, simple beneath, sparsely to ± much-branched in upper quarter, somewhat woody at base. Leaves rigid, decussate, linear to linear-lanceolate or very narrowly elliptic, 2–8 cm. long, 0.3–1 cm. wide, acute at the apex, acute to obtuse at the base, 1-nerved. Flowers 4–6-merous, heterostylous, subsessile, arranged in subsessile clusters aggregrated into terminal subglobose inflorescences up to 1.5 cm. long, 1.8 cm. wide, surrounded at the base by several reniform imbricate venose acuminate bracts ± 9 mm. wide; bracteoles oblong-spathulate, 5–6 mm. long, widened at the apex, acuminate. Calyx-tube narrowly campanulate, 3–4 mm. long, thin, 8–12-nerved; lobes triangular, 0.5 mm. long, acute, the alternating appendages erect or curved, narrow, ± 2 mm. long, ciliate above the middle. Petals purple or mauve, narrowly obovate, ± 4.5 mm. long, deciduous. Stamens as many as the calyx-lobes. Short-styled flowers: longer filaments 7 mm. long, shorter filaments 5.5 mm. long, overtopping stigma by 2.5 and 1 mm. respectively; style 4.2 mm. long. Long-styled flowers: longer filaments 6 mm. long, shorter 4 mm. long; style 7.5 mm. long. Ovary ellipsoid, 1.7 mm. long, 2-locular. Immature capsule ellipsoid-globose, 3 mm. long. Fig. 8/22–29, p. 25.

TANZANIA. Uzaramo District: 28 km. SE. of Dar es Salaam, 2 km. SE. of Fungoni Pond, 9 Sept. 1977, *Wingfield* 4167! & 55 km. S. of Dar es Salaam, Mkuranga–Kisiju road, km. 33, 15 Sept. 1977, *Wingfield* 4192!; Lindi District: Selous Game Reserve, ± 38 km. SW. of Kingupira, 5 Aug. 1975, *Vollesen in M.R.C.* 2625!
DISTR. T 6, 8; Mozambique
HAB. Seasonally waterlogged grassland (e.g. *Hyparrhenia rufa* and *Oryza longistaminata*) on blackish sandy soil with *Rhynchospora candida* and other sedges; 30–35 m.

NOTE. Koehne gives a record from "Madagascar (Lindley)" and Hiern states "also occurs in Madagascar" but H. Perrier de la Bâthie (Fl. Madag. 147, Lythracées: 9 (1954)) states it has never been refound.

10. **N. dodecandra** *(DC.)* *Koehne* in E.J. 3: 334 (1882) & in E.P. IV. 216: 234 (1903); F.W.T.A., ed. 2, 1: 166 (1954); E.P.A.: 611 (1959); M.G. Gilbert in Fl. Eth., ined. Types: Senegal [Walo, Richard-Tol], *Perrottet** 58 & s.n. (G, syn., K, ?isosyn.!, P, isosyn.)

Glabrous perennial herb 25–60 cm. tall, ± glaucous; stems 4-angled or almost winged above, somewhat branched. Leaves decussate, linear-oblong to lanceolate, 1–6.5 cm. long, 1.5–5 mm. wide, narrowly tapering-acute at the apex, truncate to subcordate at the base, sometimes ± asperulous at the margins. Flowers 5–7-merous in 1–3-flowered axillary dichasia in axils of small leaves at apices of stems; peduncles 0.15–1.1 cm. long; pedicels up to ± 1 mm. long; primary bracteoles at apex of peduncles narrowly elliptic to lanceolate, 2–6 mm. long, 0.8–1.2 mm. wide; secondary bracteoles lanceolate, up to 2 mm. long. Calyx-tube ± campanulate, sometimes ± urceolate in fruit, ± 4 mm. long, 10–14-plicate-ribbed; lobes triangular, ± 1 mm. long, the intermediate appendages 0.5 mm. long, spreading. Petals pink, obovate, 7 mm. long, 5 mm. wide. Stamens 10–14, exserted. Ovary ellipsoid, 4–5-locular; style 6–7 mm. long. Fig. 8/30–35, p. 25.

UGANDA. Karamoja District: Meris Camp, June 1930, *Liebenberg* 208!
DISTR. U 1; Senegal, Sudan and Ethiopia
HAB. Seasonal river in dry country but no data given; 1200 m.

SYN. *Ammannia dodecandra* DC. in Mém. Soc. Phys. Hist. Nat. Genève 3(2): 89, t. 2 (1826) & Prodr. 3: 80 (1828); A. Rich., Tent. Fl. Abyss. 1: 279 (1847)
      Nesaea candollei Guill. & Perr., Fl. Seneg. Tent. 1: 307 (1833); Hiern in F.T.A. 2: 473 (1871), *nom. illegit.* Type as for *Nesaea dodecandra*

NOTE. The single specimen seen from the Flora area agrees quite well with typical material; it has not previously been reported from East Africa. Richard reported stamens 8 for Quartin Dillon's Shire [Chiré] specimens.

11. **N. stuhlmannii** *Koehne* in Gilg in P.O.A. C: 286 (1895), in E.J. 22: 150 (1895) & in E.P. IV. 216: 235 (1903). Type: Tanzania, Pangani, *Stuhlmann* 516 & 297 (B, syn.†)

* F.W.T.A. cites *Perrottet* 58, 129 and 339, Koehne only 339; Kew sheet is 129; Geneva microfiche 58 and some unnumbered sheets.

Perennial glabrous herb, 0.45–1.8 m. tall, with stiffly erect pale single stems, ± woody but slender and probably from a woody rootstock, branched in upper part, the slender branchlets angular; lower parts of main stem with bark fibrillating. Leaves grey-green, stiff, erect, sessile, decussate, longer than the internodes, linear-oblong to narrowly oblong-lanceolate, 1–2.8(–4) cm. long, 1.5–5(–12) mm. wide, ± sharply acute at the apex, rounded or slightly subcordate at the base, margin recurved when dry; midrib strongly raised beneath but lateral venation invisible or very obscure. Cymes numerous, 1–3-flowered, borne on the branchlets mostly one from each axil of a leaf-pair, often deciduous leaving axes; peduncle filiform, 4–8 mm. long; pedicels 1.5–2 mm. long; bracteoles narrowly oblong, ± 1(–2) mm. long. Calyx-tube obconic to campanulate, 2.5–3 mm. long, very grooved in young flowers in dry state; lobes 5–6, broadly triangular 0.6–1 mm. long, not mucronate, with 2 impressed nerves; appendages 6, obvious in young buds but later obscure, scarcely 0.8 mm. long (subnulli *fide* Koehne). Petals 5–6, red-mauve or pink, elliptic, 4 mm. long, 3 mm. wide. Stamens 10–12; filaments 2.5 mm. long. Style at least 4–6 mm. long; stigma capitate. Fig. 9/1–8, p. 29.

KENYA. Kilifi District: 19.2 km. NW. of Malindi, May 1960, *Rawlins* 865!; Tana River District: opposite side of river to Ngao, 7 Mar. 1977, *Hooper & Townsend* 1244! & Kurawa, 21 Sept. 1961, *Polhill & Paulo* 533!
TANZANIA. Pangani District: Mwera, Mseko, Kitupa, 15 Mar. 1956, *Tanner* 2648! & Pangani, *Stuhlmann* 297, 516
DISTR. K 7; T 3; not known elsewhere
HAB. Open grassland on heavy black clay; valley with grass and *Suaeda*, *Brachystegia* woodland, irrigated rice-fields; 15–75 m.

12. **N. parkeri** *Verdc.*, sp. nov. aff. *N. stuhlmannii* Koehne habitu herbaceo praecipue annuo, foliis angustioribus linearibus vel lineari-lanceolatis, calycis tubo basi plerumque angustato obtriangulari haud conspicue plicato differt. Typus: Kenya, Kitui District, Lali Hills, Dakadima–Didima, *Parker* GM/501/H (EA, holo.!)

Usually annual unbranched to much branched ± erect often wiry herb, 15–30 cm. tall; stems pale greyish, slender but distinctly woody below with peeling bark, glabrous or slightly scabrid above on the angles. Leaves linear to linear-lanceolate, 0.4–6 cm. long, 0.5–4 mm. wide, acute at the apex, ± rounded to subauriculate at the base, glabrous. Flowers 5–6-merous, probably heterostylous; cymes numerous, borne in most axils, 1–5-flowered; peduncle filiform, 5–8(–12) mm.; pedicels 1–2 mm. long or apparently up to 4 mm. (see note); bracteoles linear-oblong, 0.5 mm. long. Calyx-tube obtriangular, 2 mm. long, narrowed at the base, minutely papillate; lobes triangular, 0.5–0.8 mm. long; appendages obsolete. Petals pinkish apricot or orange, obovate, 2.5–4 mm. long, 1.2–3 mm. wide. Stamens 10; filaments 2–3 mm. long. Ovary ellipsoid, 1.5 mm. long; style 2.5–6 mm. long. Fruit broadly ellipsoid.

var. **parkeri**

Leaves 1–3 cm. long, 0.5–4 mm. wide. Fig. 9/9–16, p. 29.

KENYA. Kitui District: Lali Hills, Dakadima–Didima, 31 Mar. 1963, *Parker* GM/501/H!; Teita District: Tsavo National Park East, Dika Plains to Dida Harca road, 3°28'S 38°52'E, 23 Jan. 1972, *R.B. & A.J. Faden* 72/109!; Tana River District: Tana River National Primate Reserve, 21 Mar. 1990, *Luke et al. T.P.R.* 761!
DISTR. K 4, 7; not known elsewhere
HAB. Edges of water-pans in open bushland, seasonal pools with sedges and grasses, scattered tree grassland and flooded grassland in forest glades; 30–460 m.

NOTE. *Luke et al. T.P.R.* 32.6 (Tana River Primate Reserve, Mehechele, 13 Mar. 1990) has apparent pedicels to 4 mm. but they are lateral axes of the cymes and not true pedicels; it is also not easy to distinguish narrow calyx-bases from pedicels.

var. **longifolia** *Verdc.*, var. nov. a var. *parkeri* foliis anguste lineari-lanceolatis usque 6 cm. longis, 4 mm. latis differt. Typus: Kenya, "north of Mombasa to Lamu and Witu", *Whyte* (K, holo.!, BM, iso.!)

Habit not known. Leaves linear-lanceolate up to 6 cm. long, 4 mm. wide.

KENYA. "North of Mombasa to Lamu and Witu", *Whyte*!
DISTR. K 7; not known elsewhere
HAB. Probably forest glades; not known

NOTE. E.G. Baker who visited Berlin with numerous queries wrote 'allied to *N. stuhlmannii*' on the BM duplicate.

GENERAL NOTE. I have decided to separate *N. parkeri* from *N. stuhlmannii* specifically after initially thinking it might be better considered a variety. Nevertheless *Luke et al. T.P.R.* 285 (Tana River Primate Reserve) is in some ways intermediate in the calyx characters and may not be an annual. Typical material of the taxa is widely different but may perhaps be solely explained by habitat differences but I suspect not.

13. **N. fruticosa** *A. Fernandes & Diniz* in Bol. Soc. Brot., sér. 2, 31: 159, t. 10 & 11 (1957) & 48: 121 (1974) & in F.Z. 4: 292 (1978). Type: Tanzania, Iringa District, by Little Ruaha R., *Semsei* 2452 (COI, holo., BR, EA, FHT, K, iso.!)

Almost glabrous much-branched shrub (0.4–)0.9–2.3 m. tall; stems 4-angled or narrowly winged. Leaves decussate, elliptic-lanceolate, 0.3–3.5 cm. long, 0.15–1.2 cm. wide, acute at the apex, subcordate at the base, entire, glossy, 1-nerved; leaves on branches much smaller than those on stem. Cymes numerous, axillary, 1–6-flowered; peduncle ± 2 mm. long; pedicels 2–5 mm. long those of pole flowers longest; bracteoles lanceolate, ± 0.5 mm. long. Flowers (4–)5(–6)-merous, probably not heterostylous. Calyx-tube oblong-ellipsoid, 3–3.5 mm. long, 8-, 10- or 12-ribbed; lobes reddish, triangular, 0.5(–0.8) mm. long, scarious at apex, the appendages very short. Petals red or pink, oblong, 3–4 mm. long, corrugated, with mid-nerve thickened beneath. Stamens twice as many as calyx-lobes, inserted near base of tube, alternate ones slightly longer and shorter, the filaments 5.3–7 mm. long, the longest opposite the sepals. Ovary ellipsoid, 2 mm. long, 1.6 mm. wide, 3-locular; style 7 mm. long overtopping the stamens. Capsule globose, ± 3 mm. diameter, included in the tube. Seeds numerous, ± 0.5 mm. long. Fig.-9/17–26.

TANZANIA. Iringa District: by Little Ruaha R., 8 Sept. 1956, *Semsei* 2452! & Great Ruaha R., Nyamakuyu Rapids, 19 Aug. 1969, *Greenway & Kanuri* 13770!
DISTR. T 7; Zambia
HAB. River banks, sand banks amongst rocks in river bed with riverine fringe of *Tamarindus, Dalbergia, Newtonia, Acacia*, etc.; 780 m.

NOTE. *Carter et al.* 2456 (Chunya District: 12 km. NW. of Saza on road to Kwimba Mt. (on rocky outcrop in miombo woodland at 1000 m.) probably belongs here but has uniformly small leaves about 4–9 mm. long.

14. **N. heptamera** *Hiern* in F.T.A. 2: 472 (1871); Koehne in E.J. 3: 335 (1882) & in E.P. IV. 216: 235 (1903); Boutique in F.C.B., Lythraceae: 20, t. 2 (1967); A. Fernandes in F.Z. 4: 291 (1978) & in Fl. Moçamb. 73: 17 (1980); Immelman in Bothalia 21: 45 (1991). Type: Malawi, 'Zomba and East-end of Lake Shirwa', *Meller* (K, holo.!)*

Erect glabrous subshrubby herb, 7–40 cm. tall, with several shoots from a woody rootstock, often with 1–15 narrow tubers to 25 cm. long; stems often reddish, narrowly winged, branched mostly near the base. Leaves decussate, sessile, lanceolate, narrowly oblong-elliptic or oblong, 0.5–5.8 cm. long, 0.08–1.2 cm. wide, acute at the apex, rounded, truncate or subcordate at the base, usually thickish and ± glaucous, 1-nerved; margins minutely serrulate. Flowers (5–)6–7(–8)-merous in (1–2)3(–7)-flowered axillary dichasial cymes; peduncles slender, 0.5–2(–3.5) cm. long; pedicels 1–2 mm. long; bracteoles linear-lanceolate to ovate-lanceolate, 0.5–2 mm. long. Calyx-tube cupular, 3–4.7 mm. long; lobes triangular, 0.8–1 mm. long, acute at the apex; appendages very short. Petals orange, vermilion red or pink, sometimes with dark midrib, obovate, 2.5–6.5 mm. long, 4 mm. wide. Stamens (10–)12–14(–16), exserted, those opposite the sepals 6–7 mm. long, the others 4–5 mm. long; filaments crimson. Ovary globose, 1–1.5 mm. in diameter; style yellow, (6–)8–9 mm. long, overtopping the anthers in long-styled flowers, 2–3 mm. long in short-styled flowers, sometimes kinked. Capsule globose, 3–3.5 mm. diameter. Seeds concavo-convex, ovoid ± 0.5 mm. long. Fig. 9/27–36.

* The label is a general printed one and it is not clear if the material came from one or both of the localities or from one or more plants; I suspect the 5 pieces (mounted on one sheet) came from one plant and have called it a holotype.

FIG. 9. *NESAEA STUHLMANNII*—1, habit, × 1; 2, flower, × 4; 3, petal, × 6; 4, part of corolla, opened out, × 10; 5, 6, stamens, × 10; 7, pistil, × 8; 8, fruit (measured dry), × 4. *N. PARKERI* var. *PARKERI* 9, habit, × 1; 10, flower, × 4; 11, petal, × 6; 12, part of corolla, opened out, × 10; 13, 14, stamens, × 10; 15, pistil, × 10; 16, fruit, × 4. *N. FRUTICOSA*—17, habit, × 1; 18, flower, × 4; 19, petal, × 6; 20, part of corolla, opened out, × 10; 21, 22, stamens, ×

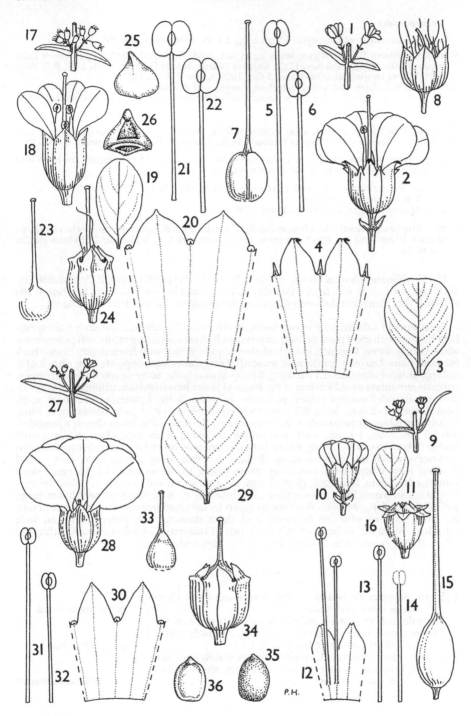

10; **23**, pistil, × 4; **24**, fruit, × 4; **25, 26**, immature seed, × 24. *N. HEPTAMERA*—**27**, habit, × ⅔; **28**, flower, × 4; **29**, petal, × 6; **30**, part of corolla, opened out, × 10; **31, 32**, stamens, × 4; **33**, pistil, × 6; **34**, fruit, × 4; **35, 36**, seed, × 24. 1–7, from *Hooper & Townsend* 1244; 8, from *Polhill & Paulo* 535; 9–16, from *Parker* GM/501/H; 17–26, from *Semsei* 2452; 27–33, from *Michelmore* 1135; 34–36, from *R.L. Davies* 156. Drawn by Pat Halliday.

var. **heptamera**

Leaves lanceolate to oblong, up to 4.5 cm. long, 1.2 cm. wide, never whorled.

TANZANIA.   Tabora District: Kigwa, Walla R., 11 Jan. 1964, *Carmichael* 1055!; Ufipa District: Lake Rukwa, 16 Dec. 1935, *Michelmore* 1135!; Kondoa District: Kikore [Kikori], 5 Feb. 1930, *B.D. Burtt* 2711!; Mbeya District: Mbimba–Mbosi, 7 Oct. 1958, *Reakes-Williams* 128!
DISTR.   **T** 1, 4–7; Zaire, Mozambique, Malawi, Zambia and Zimbabwe
HAB.   Open seasonally flooded grassy plains, *Combretum* wooded grassland with *Microchloa indica*, river flats and rice fields; 900–1260 m.

var. **bullockii** *Verdc.*, var. nov., a var. *heptamera* foliis linearibus ± 11 mm. longis, 0.8 mm. latis interdum ± verticillatis differt. Typus: Tanzania, Buha District, *Bullock* 3203 (K, holo.!, BR, iso.!)

Leaves linear, ± 11 mm. long, 0.8 mm. wide, sometimes ± in whorls of 3, occasionally much reduced above so that inflorescence looks like a raceme of cymes.

TANZANIA.   Buha District: 19.2 km. E. of Mgende, 22 Aug. 1950, *Bullock* 3203!
DISTR.   **T** 4; not known elsewhere
HAB.   Edges of 'mbugas' on hard black soil; 1200 m.

NOTE.   This has a distinctive look but can hardly be anything but *N. heptamera;* the root has ± 10 large tubers ± 10 cm. long but some specimens of var. *heptamera* (e.g. *Michelmore* 1135) have similar rootstocks.

15. **N. kilimandscharica** *Koehne* in Gilg in P.O.A. C: 286 (1895) & in E.P. IV. 216: 236, fig. 46D (1903); V.E. 3(2): 653, fig. 287D (1921); A. Fernandes in Garcia de Orta, Sér. Bot. 4: 190 (1980). Type: Tanzania, between Meru and Kilimanjaro, *Volkens* 1657 (B, holo.†)

Erect ericoid subshrub or woody herb, 18–70 cm. tall, glabrous, minutely pubescent-hirtellous or with most parts densely covered with short spreading rather stiff pubescence when young; stems 4-angled, branched above, quite woody with fibrous grey-brown bark on older parts. Leaves ± sessile, held somewhat erect, alternate, opposite or in whorls of 3, oblong-oblanceolate, 0.3–2.7 cm. long, 0.8–3(–8) mm. wide, acute or apiculate at the apex, cordate-auriculate or subhastate at the base. Flowers heterostylous, trimorphic, 4–6(–7)-merous, in 1–5-flowered cymes; peduncle 1–9 mm. long; bracteoles brown at apex, boat-shaped, 1–2 mm. long, 0.5 mm. wide; pedicels 0.5–2(–4 pole flower) mm. long. Calyx-tube tubular-campanulate, 2.5–4 mm. long, 2.5 mm. wide; lobes shortly triangular, 0.7–1.5 mm. long, 1.1 mm. wide, mucronate, reflexed; sinus-appendages minute, only visible in bud. Petals 4–7, orange to red or maroon, mostly brick-red or pale carmine with mid-nerve much darker, obovate-spathulate, 4–5 mm. long, 2 mm. wide. Long-styled flowers: long stamens 6.5 mm. long, overtopping the calyx-lobes by 3.5 mm.; short stamens 3.5–4 mm. long; style (5–)6–7 mm. long, exserted 4–6 mm. Short-styled flowers: long stamens 8 mm. long, overtopping calyx-lobes by 4.5 mm.; short stamens 5 mm. long; style 2–3 mm. long. Mesostylous flowers (seen by me) had intermediate measurements, the long stamens with red filaments and short stamens with green filaments; style cylindric, intermediate between the two kinds of stamens. Capsule included within the calyx, oblong or obovate, 3–4 mm. long, 2.1–3 mm. wide.

KEY TO VARIETIES

1. Plant glabrous or nearly so, the leaves minutely shortly
   ciliate when very young . . . .   b. var. **ngongensis**
   Plant densely to sparsely pubescent or rarely glabrescent   2
2. Leaves mostly erect, ± minutely auriculate, densely covered
   with short spreading hairs (**K** 3, 4, 6)   . . . . .   c. var. **hispidula**
   Leaves often ± horizontal, sometimes more widely auriculate,
   glabrescent to covered with fine very short hairs or
   minute pubescence . . . .   a. var. **kilimandscharica**

FIG. 10. *NESAEA KILIMANDSCHARICA* var. *NGONGENSIS*—**1**, habit, × 1; **2**, short-styled flower, × 4; **3**, petal, × 6; **4, 5**, stamens, × 4; **6**, short-styled pistil, × 4; **7**, short-styled fruit, × 4; **8, 9**, long styled stamens, × 4; **10**, long-styled pistil, × 4; **11**, long-styled fruit, × 4; **12**, seed, × 24. *N. KILIMANDSCHARICA* var. *HISPIDULA*—**13**, habit, × 1; **14**, long-styled flower, × 4; **15**, petal, × 6; **16, 17**, long-styled stamens, × 10; **18**, long-styled pistil, × 4; **19**, short-styled

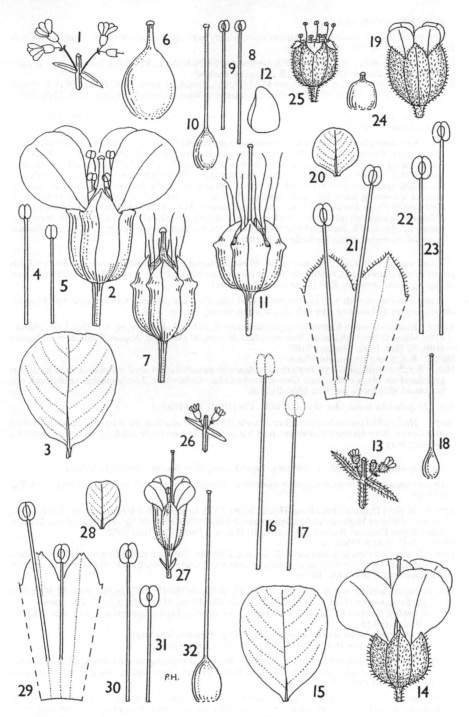

flower, × 4; **20**, petal, × 6; **21**, part of corolla, opened out, × 10; **22, 23**, short-styled stamens, × 10; **24**, short-styled pistil, × 6; **25**, short-styled fruit (measured dry), × 4. *N. SCHINZII* subsp. *SUBALATA*—**26**, habit, × 1; **27**, flower, × 4; **28**, petal, × 6; **29**, part of corolla, opened out, × 10; **30, 31**, stamens, × 10; **32**, pistil, × 8. 1–10, from *Napier* 2479; 11, 12, from *Greenway* 9168; 13–25, from *Kirrika* s.n.; 26–32, from *Eggeling* 2779. Drawn by Pat Halliday.

a. var. **kilimandscharica**

Leaves mostly ± horizontal, sometimes more widely auriculate, glabrescent to covered with very fine short hairs or minute pubescence.

KENYA. Nairobi, Athi road, 22 Mar. 1953, *Verdcourt* 919 & Nairobi, 22 May 1915, *Dummer* 1934, in part! & *Blayney Percival*! & 30 Aug. 1925, *B. de Nadamlemski* 4!

TANZANIA. Moshi District: 12.8 km. on Moshi–Arusha road, 27 June 1970, *Kabuye* 161! & Moshi, Doloti, Jan. 1928, *Haarer* 922! & Viehboma–Momela [Njoro Lkatende], 8 Nov. 1901, *Uhlig* 385!

DISTR. K 4, 6; T 2; not known elsewhere

HAB. Seasonally wet grassland and persisting on waste ground with *Tagetes, Schkuria, Orthosiphon* and rubbish; ± 900–1650 m.

NOTE. Fernandes has annotated and confirmed a specimen *F. Thomas* 119 as this which seems very close to var. *hispidula*. The duplicate was sent from Berlin and the determination is in Koehne's handwriting. It is labelled Nairobi, water ditches in grassland, 12 Apr. 1903, 1200 m. The altitude is some 350 m. lower than Nairobi so cannot be from there. There is no doubt that var. *hispidula* and var. *kilimandscharica* are very close but typical specimens of the former with erect leaves and distinct spreading hairs are easily distinguished but a good deal of further study is required. *Verdcourt & Fraser Darling* 2293 (Kenya, Masai District, Mara Plains, Keekorok [Egalok], 21 Oct. 1958) a tough-rooted subshrubby herb in grassland with scrub is related to both but the leaves are mostly in whorls of 3, the main bracteoles up to 3 mm. long and calyces shorter; without additional material no decision can be made on its status.

b. var. **ngongensis** *Verdc.*, var. nov. a var. *kilimandscharica* et var. *hispidula* planta glabra vel foliis juvenilibus minute ciliatis, habitu plerumque caespitoso differt. Typus: Kenya, Masai District, W. foot of Ngong Hills, *Greenway* 9168 (K, holo.!, BR, EA, iso.!)

Stems sometimes with ± 25 caespitose ± basal branches but possibly ± single-stemmed in some plants. Leaves glabrous or minutely ciliate when young. Fig. 10/1–12.

KENYA. Kiambu District: Kikuyu Escarpment Forest (Lari Forest Reserve, etc.), 17 Dec. 1972, *Hanson* 816!; Masai District: 48 km. on Nairobi–Magadi road, 30 Jan. 1933, *Napier* 2479!* & W. of Ngong Hills, 30 Mar. 1957, *Greenway* 9168!

DISTR. K 4, 6; not known elsewhere

HAB. Scrub on hillside not far from stream; *Acacia drepanolobium, A. seyal* woodland with *Pennisetum* grassland on black cotton soil; *Croton, Trichocladus, Acokanthera, Tarchonanthus* semi-deciduous bushland with forest patches; 1650–2130 m.

SYN. [*N. lythroides* sensu Hanid in U.K.W.F.: 154 (1974), *non* Hiern]

NOTE. *Hanson* 816 comes from a higher altitude (1830–2130 m.) than the other two specimens cited and from a more forested and wetter area, but as far as can be made out from so few specimens it belongs here.

var. **hispidula** (*Rolfe*) *Verdc.*, comb. nov. Type: Kenya, near Nairobi, *Whyte* (K, lecto.!)

Leaves mostly held erect, ± minutely auriculate, densely covered with short spreading hairs. Fig. 10/13–25.

KENYA. Nairobi District: Thika Road House, 8 Oct. 1950, *Verdcourt* 354! & Karen, *Hale* 39! & Nairobi, corner of Uhuru Highway and Langata road, 8 May 1975, *Kabuye & Ng'weno* 521!; Masai District: Mara Game Reserve, Burrungat [Burrugali] Plains, 2 Dec. 1971, *Taiti* 1862!

DISTR. K ?3, 4, 6; S. Ethiopia

HAB. Rocky outcrops in grassland with *Grewia, Lantana, Tagetes* scrub, often on shallow soil and persisting in damp places near and in cultivations and waste ground; rocky dry water courses, also on black cotton soil; 1400–1800 m.

SYN. *N. hispidula* Rolfe in K.B. 1916: 230 (1916); Heriz-Smith, Wild Fl. Nairobi Nat. Park: 26, 50 (1962); A. Fernandes in Garcia de Orta, sér Bot. 4: 190 (1980); M.G. Gilbert in Fl. Eth., ined.
      N. winkleri Koehne in V.E. 3(2): 653 (1921), *nom. invalid.* Typical specimen Kenya, near Nairobi, *Winkler* (?B,†)
      [*N. kilimandscharica* sensu Cufod., E.P.A.: 611 (1959), *non* Koehne]
      [*N. lythroides* sensu Hanid in U.K.W.F.: 154 (1974), *non* Hiern]

NOTE. Although published in a key *N. winkleri* is not actually distinguished from *N. kilimandscharica* and cannot be considered validly published. It is not clear if Fernandes, who gave the synonymy, actually saw any material.

NOTE. (on species as a whole). Many specimens of *N. kilimandscharica* have been named *N. lythroides* Hiern but although similar the latter differs in its almost hastate leaves, very numerous inflorescences, longer pubescence, different bracts and bracteoles and generally different appearance. This Angolan species is still only known from a few specimens collected by Welwitsch.

* A tiny separate label gives Dr [V.G.L.] van Someren.

16. **N. schinzii** *Koehne* in Verh. Bot. Ver. Brandenb. 30: 250 (1888) & in E.J. 22: 151 (1895) & in J.B. 40: 69 (1902) & in E.P. IV. 216: 239 (1903); Pohnert & Roessler, Prodr. Fl. SW.-Afr. 95: 7 (1966); Boutique, F.C.B. Lythraceae: 20 (1967); A. Fernandes in C.F.A. 4: 186 (1970) & F.Z. 4: 292 (1978) & in Fl. Moçamb. 73: 18 (1980); Maquet in Fl. Rwanda 2: 488, fig. 152.3 (1983); Immelman in Bothalia 21: 46 (1991). Type: South Africa, Cape Province, Upington, Oshando, *Schinz* 517 (B?, holo., BOL, K!, iso.)

Spreading or erect herb or subshrub with several stems, 6–90 cm. tall from a woody stock; stems glabrous, micropapillate or minutely pubescent, distinctly square or with angles almost winged, sometimes much-branched, when immediately above base appearing caespitose with up to ± 30 shoots. Leaves quite sessile or minutely petiolate, narrowly oblong, 5–12(–20) mm. long, 1.6–6(–9) mm. wide, rounded subacute at the apex, ± truncate to rounded subauriculate at the base, sometimes quite broadly so. Inflorescences 1–3(–5)-flowered; peduncle 0.5–3 mm. long; pedicels 0.3–1.2 mm. long or ± lacking; flowers trimorphic. Calyx-tube obconic or ellipsoid, 1–1.5(–3) mm. long, 1.2–1.8 mm. wide; lobes 4, triangular, (0.3–)1.5 mm. long, 0.5–1.5 mm. wide. Petals 4, red-brown, orange-red, deep pink or magenta, 1.5–4 mm. long, 0.8–4 mm. wide. Stamens 7–8, exserted, 4–7 mm. long. Style 3.5–4.7(–7) mm. long in long-styled flowers, 1–2(–4) mm. long in short-styled flowers, straight or bent below the stigma. Fruit ellipsoid, 1.5–3 mm. long, 0.8–2 mm. wide.

subsp. **subalata** (*Koehne*) *Verdc.*, comb. et stat. nov. Type: Tanzania, Mwanza District, Kagehi, *Fischer* 266 (B, syn.†, K, isosyn.!)

Plant often ± prostrate but sometimes erect; stems angled, almost winged. Leaves not minutely petiolate, more closely placed and less pointed. Petals usually smaller, 1.5 mm. long, 1 mm. wide. Fig. 10/26–32, p. 31.

UGANDA. Karamoja District: E. Kadam, Lalachat, Feb. 1936, *Eggeling* 2779!; Ankole District: Mbarara–Masaka road, Kaianja, 24 Aug. 1968, *Lock* 68/198!; Mbale District: between NE. foothills of Mt. Elgon and Mt. Kadam, Sebei, 12 Mar. 1958, *Symes* 317!
KENYA. NE. slopes of Aberdare Mts., 22 Sept. 1915, *Dowson* 541!; Kisumu-Londiani District: Lumbwa, Muhoroni district, Dec. 1939, *Opiko* in *Bally* 657! & Kisumu, Mar. 1969, *Tweedie* 3615!
TANZANIA. Mwanza District: Ukiriguru, 8 Feb. 1953, *Tanner* 1204! & Massanza I., Bujingwa, 20 Nov. 1951, *Tanner* 481!; Maswa District: 6 km. N. of Simba Kopjes, *H.M.H. Braun* 194!; Musoma District: Nyakoromo guard post–Ndabaka, 19 Apr. 1962, *Greenway & Watson* 10612!; Kondoa District: Kondoa road, Mrijo [Murijo] village, 22 Jan. 1973, *Richards* 28357!
DISTR. U 1–3; K 3, 5, 6; T 1, 2, 5; Rwanda and E. Zaire
HAB. Open grassland (e.g. *Themeda, Pennisetum, Echinochloa* with *Cyperus* and *Aeschynomene*) or grassland with scattered trees, light *Acacia* scrub, etc., and associated roadsides, usually on black cotton soil or very shallow soil on calcareous hardpan; also recorded from rock crevices; 1080–2100 m.
SYN. *N. schinzii* Koehne var. *subalata* Koehne in E.J. 22: 151 (1895) & in J.B. 40: 69 (1902) & in E.P. IV. 216: 240 (1903)
[*N. schinzii* sensu Gilg & Koehne in P.O.A. C: 286 (1895) & Boutique, F.C.B., Lythraceae: 20 (1967), *non* Koehne sensu stricto]
NOTE. Despite an instinctive feeling that this is not conspecific with typical *N. schinzii* no really valid characters have been found to support treating it as a separate species; subspecific rank has been used for purely geographical reasons the two areas of distribution being widely separated. Typical *N. schinzii* occurs in Angola, Namibia, NW. Botswana and South Africa (Transvaal and Orange Free State); Immelman records it from Zimbabwe and also mentions it for Zaire. She does not distinguish var. *subalata* but does not include it in synonymy and it appears she considers the typical taxon to occur in East Africa. Some material from SW. Africa has leaves up to 3 × 1 cm. and peduncles to 7 mm. long and pedicels to 3 mm. long. S. African material at Kew seems very heterogeneous and Immelman does not cite material in detail.

17. **N. volkensii** *Koehne* in Gilg in P.O.A. C: 286 (1895) & in E.P. IV. 216: 236 (1903). Type: Tanzania, Pare District, Muanamata, *Volkens* 2390 (B, holo.†)

Probably subshrubby; stems pale, ± 4-angled, glabrous. Leaves oblong-lanceolate or narrowly triangular-lanceolate, 1.5–4.8 cm. long, 0.5–1.2 cm. wide, attenuate-acute at the apex, the actual tip acute, dilated truncate or subcordate at the base, scabridulous-puberulous on upper revolute margins but otherwise glabrous. Flowers 6-merous, dimorphic, in paired cymes in the upper axils, several-flowered, dense or almost capitulate (*fide* Koehne) but ± loose in specimen seen; peduncles 0.9–1.5 cm. long, often connate for 2–3 mm. at the base; secondary axes ± 5 mm. long; bracteoles lanceolate, 1.5 mm. long; true pedicels 1–2 mm. long; all axes scabridulous-puberulous. Calyx-tube

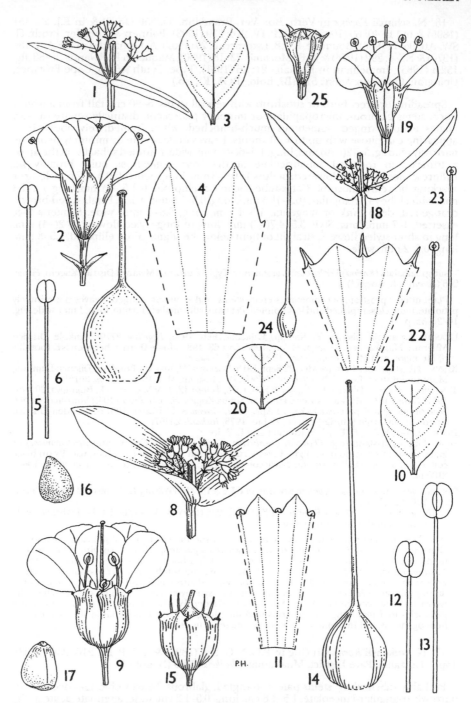

Fig. 11. *NESAEA VOLKENSII*—**1**, habit, × ⅔; **2**, flower, × 4; **3**, petal, × 6; **4**, part of corolla, opened out, × 10; **5, 6**, stamens, × 10; **7**, pistil, × 10. *N. MAXIMA*—**8**, habit, × ⅔; **9**, flower, front petal removed, × 4; **10**, petal, ×6; **11**, part of corolla, opened out, × 10; **12, 13**, stamens, × 10; **14**, pistil, × 10; **15**, fruit, × 4; **16, 17**, seed, × 24. *N. AURITA*—**18**, habit, × ⅔; **19**, flower, × 4; **20**, petal, × 6; **21**, part of corolla, opened out, × 10; **22, 23**, stamens, × 10; **24**, pistil, × 6; **25**, fruit, × 4. 1–7, from *Haarer* 1344; 8–17, from *Ludanga* in *M.R.C.* 1424; 18–25, from *Vollesen* in *M.R.C.* 2459. Drawn by Pat Halliday.

campanulate, 3.5–4 mm. long; lobes triangular, ± 1.2 mm. long, the appendages calliform thickenings. Petals obovate, 5 mm. long, 3 mm. wide. Stamens 12, slightly unequal, almost ½-exserted in short-styled flowers; ovary ellipsoid, 2.7 mm. long, 1.5 mm. wide; style ± 2.5 mm. long, just exceeding calyx-lobes (by about 1.5 mm. in specimen seen). Long-styled flowers not seen. Fig. 11/1–7.

TANZANIA. Pare District: N. Pare Mts., ± 15 km. W. of southern tip of L. Jipe, Muanamata, 27 Mar. 1893, *Volkens* 2390 & Ngulu, May 1928, *Haarer* 1344!
DISTR. **T** 3; not known elsewhere
HAB. Not known; 750 m.
NOTE. Koehne spells the type locality 'Muanamuta' but Volkens on the map in his Der Kilimandscharo (1897) and also in the text uses the spelling given above.

18. **N. maxima** *Koehne* in E.J. 41: 100 (1907) & V.E. 3(2): 653 (1921); A. Fernandes in Bol. Soc. Brot., sér. 2, 52: 9, t. 3 (1978); Vollesen in Opera Bot. 59: 52 (1980). Types: Tanzania, Uzaramo District, near Bagamoyo, Mtoni, *Stuhlmann* 7196 & "elsewhere", *Stuhlmann* 6438 (B, syn.†); Kilwa District, Selous Game Reserve, Mwendi [Mwende], *Ludanga* in *M.R.C.* 1424 (DSM, neo., EA, K, isoneo.!)

Perennial glabrous herb, 0.75–1 m. tall; stems 4-angled. Leaves decussate, sessile, elliptic (near base of stem) or narrowly oblong to lanceolate, 4–11 cm. long, 0.7–3 cm. wide, narrowly acute at the apex, subcordate to distinctly cordate at the base, ± amplexicaul; lateral nerves distinct, a basal pair forming a distinctive submarginal nerve. Flowers 6(–7)-merous, heterostylous, in dense ± 20-flowered dichasia, usually with additional branch from the base of main rhachis; apparent peduncles up to 4 mm. long but inflorescence as a whole basically sessile; pedicels up to 4 mm. long; bracteoles 0.5 mm. long. Calyx-tube campanulate, 2.5–3 mm. long, 12-nerved; lobes triangular, 0.5 mm. long, 2 mm. wide with alternating appendages inconspicuous, 0.25 mm. long. Petals brick-red or white, obovate, 4 mm. long, 3 mm. wide, clawed, corrugated, with distinct mid-nerve. Stamens twice as many as calyx-lobes. Long-styled flowers with longer stamens semi-exserted, the shorter almost ⅓ shorter. Style overtopping the stamens by ± 2 mm. Short-styled flowers with stamens almost equal, the longer 6 mm. long and shorter 5.5 mm. long, both overtopping the 3 mm. long style. Ovary obovoid-globose. Capsule globose, ± 3.5 mm. diameter, included or slightly exserted. Seeds brown, concavo-convex, ± 0.4 mm. in diameter. Fig. 11/8–17.

TANZANIA. Uzaramo District: near Bagamoyo, Mtoni, Jan. 1894, *Stuhlmann* 6438; Kilwa District: Kingupira, Lungonya Plain, 24 Feb. 1976, *Vollesen* in *M.R.C.* 3219! & Mwende, 22 June 1972, *Ludanga* in *M.R.C.* 1424! & Selous Game Reserve, *Rodgers* in *M.R.C.* 121!
DISTR. **T** 6, 8; not known elsewhere
HAB. Grassland on black clay with *Setaria, Echinochloa* and *Ischaemum*; ?50–150 m.

NOTE. It is curious this was not dealt with in the Pflanzenreich and I can only assume the material was misplaced in Berlin for a decade or so. E.G. Baker's extremely sparse sketch of the type made in Berlin (BM!) confirms the identity of this species with the material I have cited.

19. **N. aurita** *Koehne* in E.J. 41: 101 (1907); A. Fernandes in Bol. Soc. Brot., sér. 2, 49: 10, t. 2 (1975) & in 52: 10, t. 4 (1978); Vollesen in Opera Bot. 59: 52 (1980). Type: Tanzania, Lindi District, near Seliman-Mamba, Ngwai-Quelle, *Busse* 2798 (B, holo.†, BR, lecto.!, EA, isolecto.!)

Glabrous annual herb, 30–90 cm. tall; stems often purplish, 4-angled, the branches long and ascending. Leaves decussate, sessile, linear-lanceolate, 5–15 cm. long, 0.4–1.2 cm. wide, narrowly acute at the apex, narrowed to a dilated auriculate-cordate base; lateral nervation obscure. Flowers 4-merous, probably dimorphic, in numerous few-many-flowered fairly condensed dichasia up to 1.2 cm. long, 1.6 cm. wide at most axils or on some short shoots solitary or in 2-flowered cymes in axils of reduced leaves; peduncles usually very short; pedicels short with minute linear bracteoles at base or middle. Calyx-tube obconic, 1.2–2.5 mm. long, becoming cyathiform in fruit, 8-ribbed; lobes broadly triangular, 0.5 mm. long, shortly acuminate, the alternating appendages small, triangular-subulate, 0.8 mm. long, spreading. Petals violet or dark lilac, round, 1.5–2.3 mm. long. Stamens 8, inserted at middle of tube, the longer ± exserted, the shorter about ¾ as long. Ovary ellipsoid, ± 1.5 mm. long, 2-locular; style arcuate, 4–5 mm. long, much overtopping the stamens. Capsule equalling the calyx, opening by deciduous style-bearing operculum. Fig. 11/18–25.

TANZANIA. Ulanga District: between Mahenge Plateau and junction of Kilombero and Luwegu Rivers with the Rufiji, Madi, 10 June 1932, *Schlieben* 2301!; Lindi District: Nachingwea, 27 July 1952, *Anderson* 781! & Tendaguru, 19 May 1930, *Migeod* 724! & Kingupira, 20 June 1975, *Vollesen* in *M.R.C.* 2459!
DISTR. T 6, 8; not known elsewhere
HAB. Moist grassland, e.g. floodplains on black clay with *Setaria, Echinochloa* etc., or swampy areas in wooded grassland, drainage lines in *Brachystegia* woodland; 100–450 m.

SYN. *Ammannia auriculata* Willd. var. *elata* (A. Fernandes) A. Fernandes forma *longistaminata* A. Fernandes in Bol. Soc. Brot., sér. II. 52: 2 (1978). Type: Tanzania, Lindi District, Selous Game Reserve, Kingupira, *Vollesen* in *M.R.C.* 2459 (DSM, holo, EA, K, iso.!)

NOTE. There is no doubt this is close to *Ammannia auriculata* differing in little but the long style. *Beecher* 14 (Tanzania, Masasi, Aug. 1965) emphasises the problem; it consists of 3 short specimens, a specimen I have identified as *N. aurita* with style 3 mm. long; another very similar which appears to be *auriculata* (one might suspect dimorphism but *Ammannia* is never dimorphic) and a third which seems to be *Ammannia senegalensis*. Much might be learnt from a detailed study of these mixed populations in the field. If *N. aurita* is just an exceptionally long-styled variant of *A. auriculata* then it is one with a very restricted geographical distribution. There seems no point in altering the name until much more work has been done.

DOUBTFUL SPECIES

N. jaegeri *Koehne* in V.E. 3(2): 653 (1921), *nom. invalid.* Typical specimen: Tanzania, Rift valley ["Gebiet des Ostafrikanischen Grabens"], *Jaeger* (B†)

NOTE. Although this appears in a key, which at that date would be adequate to validate a name, it appears in the same couplet as *N. dinteri* with nothing but geography to distinguish it. Under the circumstances it cannot be regarded as validly published. The various key couplets leading to the species give some idea of its structure:- style mostly much longer than the ovary; calyx-appendages absent or almost so; flowers 4-merous; stamens 4; annual or biennial; stamens in front of the calyx-lobes; glabrous plant. Koehne places it in his section *Salicariastrum* and mentions it is a small swamp plant.

7. AMMANNIA

L., Sp. Pl.: 119 (1753) & Gen. Pl., ed. 5: 55 (1754); Koehne in E.P. IV. 216: 42 (1903); S.A. Graham in Journ. Arn. Arb. 66: 395–420 (1985).

Annual or possibly short-lived perennial erect or decumbent herbs or often small ephemerals, glabrous or with stem-angles, etc., with small acute emergences but mostly not hairy; stems simple or branched, 4-angled or 4-winged. Leaves decussate, sessile, linear to lanceolate or narrowly elliptic, attenuate to auriculate or cordate at the base, 1-nerved. Flowers red or greenish, 4–(5–8)-merous, monomorphic, in (1–)3-many-flowered subsessile to pedunculate axillary dichasia; pedicels developed or ± absent; bracteoles whitish, small, linear. Calyx-tube campanulate to urceolate, 8-ribbed; calyx-lobes 4(–5), triangular, mostly short and broad; appendages thick, shorter than or ± equalling the lobes or absent. Petals lacking or white, pink or purple, 1–4 (or 4–8 *fide* Fernandes), small, very deciduous. Stamens 4–8, rarely fewer, included to exserted. Ovary incompletely 2–4(–5)-locular, the septae interrupted above the placentas (or in one extra-African species unilocular with parietal placentation, but see note); ovules numerous. Style lacking or well developed, not continuous with the placentas (but see note); stigma capitate. Capsule globose, ellipsoid or depressed, thin, ± circumscissile or bursting irregularly; wall not closely striate. Seeds brown, numerous, small, obovoid, concave-convex, angular.

About 25 species of aquatic or semiaquatic herbs in both tropics and temperate parts of both Old and New Worlds.

Some species are well defined but many are very difficult. Koehne dealt with the variation in a very artificial way with numerous infraspecific taxa at various ranks mostly not correlated with geography. Collectors having the opportunity to study populations should note variation in presence, number and size of petals and also confirm that the flowers are always monomorphic as stated; any variation in style-length should be noted. I have always found the species difficult to name and am very dissatisfied with this treatment; difficulties are mentioned in the notes which field work could perhaps resolve. Hybridization should be investigated. The petals are often extremely deciduous and to be sure they are absent a bud should be dissected.

The technical differences, ovary dissepiments, lack of continuation between style and placenta used to distinguish *Ammannia* from *Nesaea* are difficult or impossible to see and Koehne's wide separation of the two is artificial; rather I would query their distinctness. This is confirmed by the work of S. A. Graham et al. mentioned in the introduction.

1. Style lacking or very short, up to ± 0.3 mm. long; petals
   often absent 2
   Style at least and mostly over 0.5 mm. long; petals mostly
   present 6
2. Calyx densely shortly papillate-pubescent (**K** 7) . . . 9. *A. urceolata*
   Calyx glabrous 3
3. Inflorescences slightly lax, the pedicels evident; petals
   absent; leaves attenuate to the base; calyx without
   appendages between the lobes . . . . 8. *A. baccifera*
   Inflorescences congested glomerules or slightly to
   distinctly lax; petals often present; leaves mostly
   subcordate, subhastate or auriculate at the base; calyx
   with or without small to distinct appendages 4
4. Inflorescences lax or at least pedicels easily evident 5
   Inflorescences sessile and condensed, the pedicels not or
   scarcely evident . . . . 7. *A. wormskioldii*
5. Petals white, linear or subulate; **T** 8 . . . . 5. *A. linearipetala*
   Petals obovate but often lacking; **K** 7, **T** 6, **Z** (coastal) 6. *A. senegalensis*
6. Style mostly longer, 0.5–2.5 mm. long; inflorescence
   condensed to very lax with peduncle up to 1.8 cm. long
   but short in some variants; capsule 1.5–3 mm. wide 7
   Style mostly shorter, 0.5–1 mm. long; inflorescences
   condensed but not sessile, the peduncle 1.5–3 mm.
   long; capsule usually small, 1.5–1.8 mm. wide save in
   A. elata known from 2 specimens 8
7. Inflorescences usually lax with peduncles 0.4–1.8 cm. long
   but forms with very short peduncles exist; style 0.5–2.5
   mm. long; capsule usually under 3 mm. wide, usually
   less exserted; very widespread and variable . . . 1. *A. auriculata*
   Inflorescences condensed with peduncle ± 3 mm. long;
   style 1.5–2 mm. long; capsule up to 3 mm. wide, well
   exserted from bowl-shaped calyx; **T** 7, Iringa, Ruaha 2. *A. sp. A*
8. Fruiting inflorescences usually under 8 mm. wide; capsule
   mostly small, 1.5–1.8 mm. wide; widespread . . . 3. *A. prieuriana*
   Fruiting inflorescences usually over 1 cm. wide; capsule
   2.5–3 mm. wide; succulent-stemmed plant known from
   one locality in Flora area, **K** 7, Teita . . . . . . 4. *A. elata*

1. **A. auriculata** *Willd.*, Hort. Berol. 1: 7, t. 7 (1803); DC., Prodr. 3: 80 (1828); Koehne in E.J. 1: 244 (1880) & E.J. 4: 389 (1883); Koehne & Gilg in P.O.A. C: 285 (1895); Koehne in E. & P. Pf. 3, 7: 7 (1898) & in E.P. IV. 216: 45, fig. 5B (1903); V.E. 3(2): 645 (1921); F.W.T.A., ed. 2, 1: 164, fig. 61 (1954); A. Fernandes & Diniz in Garcia de Orta 4: 404 (1956); Pohnert & Roessler in Prodr. Fl. SW.-Afr. 95: 2 (1966); Boutique in F.C.B., Lythraceae: 22 (1967); A. Fernandes in C.F.A. 4: 175 (1970) & in Bol. Soc. Brot., sér. 2, 52: 1 (1978) & in F.Z. 4: 305 (1978); & in Fl. Moçamb. 73: 37 (1980); S.A. Graham in Journ. Arn. Arb. 66: 403 (1985); Immelman in Bothalia 21: 41 (1991). Type: Egypt, near Rosetta, cult. in Berlin (B-WILLD 3081, lecto., microfiche!)*

Annual erect simple or ± branched herb, 10–65(–80) cm. tall; branches 4-angled and narrowly winged above, sometimes ± scabrous. Leaves linear to ± lanceolate, 1.5–8 cm. long, 0.2–1.4 cm. wide, narrowed to subacute apex, auriculate-cordate at the base or cuneate in some of the lower leaves, entire, 1-nerved, sometimes minutely scabrous. Flowers 4-merous in lax (2–)3–15-flowered dichasia; peduncle 0.4–1(–1.8) cm. long;

* Immelman gives Willdenow's t. 7 as an iconotype but makes no mention of the specimen in Willdenow's herbarium which was chosen by Graham as lectotype in 1985.

pedicels 0.5–3(–6) mm. long; bracteoles lanceolate, ± 1 mm. long. Calyx-tube campanulate, 1–1.5(–3) mm. long; lobes 4, triangular, 0.4–0.7 mm. long; appendages up to 0.3(–1) mm. long. Petals pink to mauve, ± round to obovate, 0.6–0.7(–2) mm. long and wide. Stamens 4–8, exserted; filaments 0.8–2.5 mm. long. Ovary globose, 0.8–1.5 mm. in diameter; style filiform, 0.5–2.5 mm. long. Capsule globose, (1.5–)1.7–3.5 mm. in diameter (usually under 3 in East Africa), equalling or slightly exceeding the lobes. Seeds brownish, concave-convex, 0.4–1 mm. long. Fig. 12/1–4.

UGANDA. W. Ankole, 6 Sept. 1905, *Dawe* 370!; Teso District: Serere,? Atira, July 1926, *Maitland* 1336!; Mengo District: Masaka road, Katonga Channel, 15 Sept. 1961, *Rose* 10071!

KENYA. Northern Frontier Province: Sololo, 3 Aug. 1952, *Gillett* 13689!; 29 km. SSW. of Embu, 20 Feb. 1957, *Bogdan* 4448!; Lamu District: Witu, *F. Thomas* 70!

TANZANIA. Lushoto District: Kitivo Forest Reserve, Aug. 1955, *Semsei* 2291!; Ufipa District: Milepa, 27 Apr. 1951, *Burnett* 15!; Iringa District: Ruaha National Park, banks of Ruaha R. opposite Lunda, 7 May 1968, *Renvoize & Abdallah* 2206!; Zanzibar I., Mahonda, 15 Aug. 1963, *Faulkner* 3246!

DISTR. **U** 1–4; **K** 1, 3 (see note) 4, 6, 7; **T** 1–7; **Z**; **P**; very widely distributed in Africa, Asia to India and China, Australia, W. Indies, N., C. and S. America

HAB. Mud by dams and ponds, wooded riverine flats, damp grassland, waterlogged areas on black clay soils, weed in rice in irrigated plots on black clay, damp mud and sand in disturbed ground; 0–1530 m. (see note)

SYN. *A. arenaria* Kunth, Nov. Gen. Sp. 6: 190 (1824). Type: Venezuela, Caracas, near San Fernando, *Humboldt & Bonpland* (P, holo.)

A. senegalensis Lam. var. *brasiliensis* A. St.-Hil., Fl. Brasil. Merid. 3: 135, t. 187 (1853). Type: Brazil, Minas Novas, S. Miguel, Jiquitinhonha R., collector not stated, ? *St.-Hilaire* (P, holo.)

A. auriculata Willd. var. *arenaria* (Kunth) Koehne in E.J. 1: 245 (1880) & in E.P. IV. 216: 46 (1903)

A. auriculata Willd. var. *arenaria* (Kunth) Koehne forma *brasiliensis* (A. St.-Hil.) Koehne in E.J. 1: 245 (1880) & in E.P. IV. 216: 46 (1903)

[*A. senegalensis* sensu Hutch. & Dalz., F.W.T.A. 1: 144, pro parte, & fig. 57 (1927), *non* Lam.]

Nesaea dinteri Koehne subsp. *elata* A. Fernandes in Bol. Soc. Brot., sér. 2, 48: 122, t. 9 (1974) & in F.Z. 4: 294, t. 73 (1978); Immelman in Bothalia 21: 45 (1991). Type: Zambia, Kafue National Park, Mumbwa, Chunga, *Mitchell* 18/50 (LISC, holo., COI, PRE, SRGH, iso.)

Ammannia auriculata Willd. var. *elata* (A. Fernandes) A. Fernandes in Bol. Soc. Brot., sér. 2, 52: 2 (1978)

NOTE. A very variable species. Small ephemeral forms can be distinctive, e.g. *Verdcourt* 3250 (Nairobi National Park, 21 Jan. 1962), 6–10 cm. tall and relatively unbranched. *Gilbert & Tadessa* 6543 (Mt. Elgon, 5 km. below Mt. Elgon Lodge, 9 Oct. 1981), 3–5 cm. tall growing in shallow soil on flat rocks in grassland, is from a high altitude, 1965 m., but is probably this species. Petals appear to be absent in some Pemba specimens named var. *arenaria* by Brenan. Graham gives complete American synonymy. Fernandes cited *Lea* 36 (Tanzania, Lake Rukwa flood plain, Milepa, 21 Apr. 1936) as *Nesaea dinteri* subsp. *elata* and Immelman has annotated a Kew sheet of this number similarly; she has maintained subsp. *elata* in *Nesaea dinteri* not commenting on Fernandes' later treatment. I have not kept var. *elata* up but the calyx appendages appear more distinct than is usual in this species.

2. **A. sp. A**

Strictly erect branched or unbranched glabrous annual herb, 17–34 cm. tall; stems quadrangular, almost winged at apices. Leaves lanceolate, 2–4.5 cm. long, 3–7 mm. wide, attenuate to a narrowly rounded apex, cordate and auriculate to subhastate at the base, sessile. Flowers reddish green, 5-merous, in 1–7-flowered condensed glomerules up to ± 7 mm. long in fruit; peduncle 0.5–3 mm. long; pedicels 0.5–1.5 mm. long; bracteoles oblong-elliptic, 0.6 mm. long, 0.3 mm. wide, soon deciduous. Calyx-tube bowl-shaped, 0.5–1.2 mm. long, 1–3.2 mm. wide, micropapillate; nerves wide; lobes rounded-triangular, 0.5–0.8 mm. long, 1–1.8 mm. wide; appendages absent. Petals ? purple, 1 mm. long, 0.7 mm. wide, falling almost immediately buds open. Stamens ?8; filaments 2.5 mm. long. Ovary 1 mm. in diameter; style 1.5–2 mm. long. Capsule globose, 2.5–3 mm. wide, well exserted from calyx. Seeds red, numerous, trigonous, concave-convex, ± 0.4 mm. long. Fig. 12/5–8.

TANZANIA. Iringa District: Msembe, by Great Ruaha R., 9 Dec. 1962, *Richards* 17362! & E. Ruaha Nat. Park, Mwagusi sand-river, 4 Aug. 1970, *Thulin & Mhoro* 575 B!

FIG. 12. *AMMANNIA AURICULATA*—**1**, habit, × ⅔; **2**, leaf, × ⅔; **3**, flower, × 6; **4**, fruit, × 6. *A. SP. A*—**5**, habit, × ⅔; **6**, leaf, × ⅔; **7**, flower, × 6; **8**, fruit, × 6. *A. ELATA*—**9**, habit, × ⅔; **10**, leaf, × ⅔; **11**, flower, × 6; **12**, fruit, × 6. *A. LINEARIPETALA*—**13**, habit, × ⅔; **14**, leaf, × ⅔; **15**, flower, × 8; **16**, petal, × 18; **17**, fruit, × 8. *A. SENEGALENSIS*—**18**, habit, × ⅔; **19**, leaf, × 2; **20**, flower, × 8; **21**, fruit, × 8. *A. WORMSKIOLDII*—**22**, habit, × ⅔; **23**, leaf, × ⅔; **24**, flower, × 8; **25**, fruit, × 8. *A. BACCIFERA*—**26**, habit, × ⅔; **27**, leaf, × ⅔; **28**,

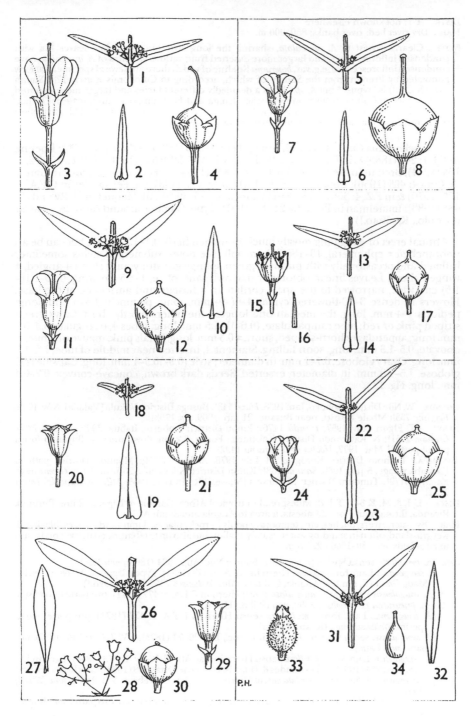

inflorescence (diagrammatic), × 4; **29**, flower, × 10; **30**, fruit, × 6. *A. URCEOLATA*—**31**, habit, × ⅔; **32**, leaf, × ⅔; **33**, flower, × 4; **34**, ovary, × 6. 1–4, from *F. Thomas* 70; 5–8, from *Richards* 17862; 9–12, from *Bally* 1270; 13–17, from *Milne-Redhead & Taylor* 9239; 18–21, from *Wingfield* 1745; 22–25, from *Norman* 247; 26–30, from *Greenway & Kanuri* 12534; 31–34, from *Luke et al.* TPR 780. Drawn by Pat Halliday.

DISTR. **T** 7; not known elsewhere
HAB. Dry river bed; river banks; 850–900 m.

NOTE. Clearly closest to *A. auriculata*, sharing the long style and auriculate leaves but with condensed inflorescences, and larger more exserted fruit; various varieties of *A. auriculata* have condensed inflorescences, e.g. var. *bojeriana* Koehne of which there are several specimens at Kew annotated by Brenan from the Sudan, of which, according to Cufodontis *Kotschy* 178 (Sudan Arasch-Cool) is "typus"* but *A. sp.* A has a decidedly different facies and larger more exserted fruits. I have refrained from naming it since Iringa and Fl. Zambesiaca area material of lax-flowered typical *A. auriculata* also has something of this facies hinting at a complicated reticulation of characters. It may after all be only a marked variant of *A. auriculata*.

3. **A. prieuriana** *Guill. & Perr.*, Fl. Seneg. Tent. 1: 303 (1833), as *"prieureana"**; Koehne in E.J. 1: 248 (1880) & in E.P. IV. 216: 48 (1903) & in V.E. 3(2): 646 (1921); F.W.T.A., ed. 2, 1: 164 (1954); Brenan in Mem. N.Y. Bot. Gard. 8: 441 (1954); A. Fernandes & Diniz in Garcia de Orta 4: 405 (1956); Boutique, F.C.B., Lythraceae: 23 (1967); A. Fernandes in C.F.A. 4: 175 (1970) & in F.Z. 4: 306 (1978) & in Fl. Moçamb. 73: 39 (1980); Maquet in Fl. Rwanda 2: 488 (1983); Immelman in Bothalia 21: 40 (1991). Type: Gambia, around Albreda, *Leprieur* (G, holo., K, photo.!)

Annual erect or straggling mostly much-branched herb, 0.1–1.2 m. tall, but can be an unbranched ± ephemeral, 10–18(–25) cm. tall (see note); submerged stems sometimes inflated and occasionally with purplish green spongy growth; young stems 4-angled, ± winged above. Leaves linear, oblong or narrowly lanceolate, 1–8.5 cm. long, 0.15–1.1(–1.5) cm. wide, narrowed to the apex, cordate to ± hastate and auriculate at the base. Flowers in dense 3–20-flowered cymes 4–6(–9) mm. long; peduncle 1.5–3 mm. long; pedicels 1–4 mm. long, the median one longest, the others mostly short. Calyx green striped pink, or red; tube campanulate, (0.6–)1–1.5 mm. long; lobes 4, triangular, 0.3–0.7 mm. long; appendages horn-shaped, short, ± 0.3 mm. long. Petals pink, mauve or white, 4, obovate, 0.5–1.3 mm. long, soon falling. Stamens 4, inserted near middle of tube, 1.2–1.5 mm. long. Ovary globose, 0.7–1 mm. in diameter; style (0.3–)0.5–1 mm. long. Capsule red, globose, 1.5–1.8 mm. in diameter, exserted. Seeds dark brown, concave-convex, 0.2–0.4 mm. long. Fig. 13.

UGANDA. W. Nile District: Koboko, Jan. 1938, *Hazel* 543!; Busoga District: Vukula [Vakula], Nov. 1932, *Eggeling* 733!; Mbale District: near Busano, 21 Jan. 1969, *Lye* 1703!
KENYA. NE. Elgon, Nov. 1957, *Tweedie* 1476!; Embu District: Emberre, Itabua, 29 Sept. 1932, *M.D. Graham* 2251!; N. Kavirondo District: Kakamega Forest, 3.2 km. down track to Buyango forest patrol post, 21 Mar. 1977, *Hooper & Townsend* 1492!
TANZANIA. Lushoto District: Korogwe, 21 July 1978, *Archbold* 2374!; Mpanda District: path to Kasimba village, 5 July 1968, *Sanane* 208!; Kilosa District: Kikwaraza, 20 June 1973, *Greenway & Kanuri* 15179!; Tunduru District: Muhuwesi [Mawesi] R., 18 Dec. 1955, *Milne-Redhead & Taylor* 7711!
DISTR. U 1, 3, ?4; K 3–7; T 1–8; widespread in tropical Africa, Gambia to Nigeria, Zaire, Burundi, Rwanda, throughout Flora Zambesiaca area to Angola and South Africa
HAB. By water, rivers, ditches, etc., temporary pools in rocky places, damp sandy places, black soil, wet grassland, often in standing water usually shallow but also up to 60 cm. deep in river beds, also in rice-fields, etc.; 30–1800(–2250) m.

SYN. [*A. vesicatoria* sensu Speke, Journ. Disc. Source Nile, App.: 634 (1863), *non* Roxb.]
A. senegalensis Lam. forma *patens* Hiern in F.T.A. 2: 477 (1871). Types: Uganda, W. Nile District, Madi, *Grant* (K, syn.!) & Angola, Golungo Alto, *Welwitsch* 2352 (BM, isosyn.!)
[*A. senegalensis* Lam. forma *auriculata* sensu Hiern in F.T.A. 2: 477 (1871), pro parte quoad syn. *A. prieuriana* et *Welwitsch* 2350, *non* (Willd.) Hiern]
[*A. senegalensis* Lam. forma *multiflora* sensu Hiern in F.T.A. 2: 477 (1871), pro parte quoad specim. *Welwitsch*, *non* (Roxb.) Hiern]
[*A. senegalensis* sensu Oliv. in Trans. Linn. Soc., Bot. 29: 74 (1873); F.W.T.A. 1: 144 (1927), pro parte, *non* Lam.]
[*A. senegalensis* Lam. var. *multiflora* sensu Hiern in Cat. Afr. Pl. Welw. 1: 373 (1898); Exell in J.B. 71, Suppl.: 234 (1933); A. Fernandes & Diniz in Garcia de Orta 6: 98 (1958), *non* (Roxb.) Hiern]
[*A. senegalensis* Lam. var. *auriculata* sensu Hiern in Cat. Afr. Pl. Welw. 1: 373 (1898), *non* (Willd.) Hiern]
[*A. auriculata* Willd. var. *arenaria* sensu Koehne in E.P. IV. 216: 46 (1903), pro parte, *non* (Kunth) Koehne]

---

* This is correct (although actually a syntype - Kew sheet will be an isosyntype) since although Koehne cites no collectors, only localities, he cites *Kotschy* 178 in his list of exsiccatae as this variety.
** Also so spelt by several of the authors cited.

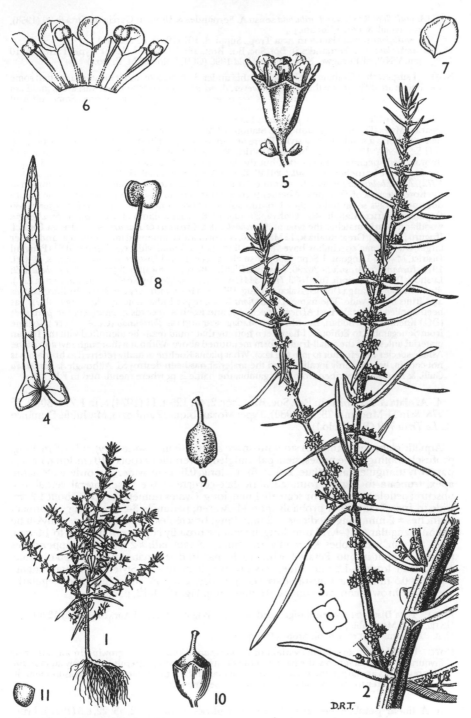

FIG. 13. *AMMANNIA PRIEURIANA*—**1**, habit, × ¹⁄₆; **2**, habit, × 1; **3**, cross-section of stem, × 2; **4**, leaf, × 3; **5**, flower, × 12; **6**, flower, spread out, × 12; **7**, petal, × 18; **8**, stamen, × 18; **9**, gynoecium, × 12; **10**, fruit, × 12; **11**, seed, × 18. Material from which plate was drawn not recorded. Drawn by D.R. Thompson with some alteration of gynoecium and fruit by Pat Halliday.

[*A. multiflora* Roxb. var. *floribunda* sensu A. Fernandes & Diniz in Garcia de Orta 6: 99 (1958), *non* (Guill. & Perr.) Koehne]
[*A. multiflora* sensu Haerdi in Acta Trop., Suppl. 8: 161 (1964), *non* Roxb.]
A. archboldiana A. Fernandes in Bol. Soc. Bot. Brot., sér 2, 52: 3, t. 1 (1978). Type: Tanzania, 10 km. WNW. of Korogwe, Kwa Sigi, *Archbold* 1996 (COI, holo., DSM, K, iso.!)

NOTE. I suspect that Graham might include this under *A. auriculata* and it must be admitted some specimens are difficult; on the whole, however, *A. prieuriana* is constant over vast areas and I am content to follow Brenan's concept as represented by his determinations. Some difficult specimens might be hybrids, e.g. *R.A. Nicholson* 243 (48 km. from Mbeya on Iringa road, Sonyanga, 1 May 1972) which resembles *A. prieuriana* but has a style 1.5 mm. long; *Bogdan* 4615 (Kenya, Embu District, Mwea-Tebere Irrigation Expt. Station, 12 Aug. 1958, a weed in irrigated plots on black heavy soil) has a similarly long style. There are undoubtedly unbranched ephemeral forms, e.g. *G.H.S. Wood* Y 13 (Uganda, Busoga District, Buswale, Iqui, 18 Oct. 1950) which consists of unbranched specimens 18 cm. tall, but the label mentions it can grow to 60 cm. and larger specimens are branched, and *Gillett* 20149 (Kenya, Meru National Park, Kiolu R. crossing, 25 Dec. 1972). *A. archboldiana* seems to be no more than such a form; large specimens are also common in the Korogwe area of Tanzania. Whether there is some genetic basis needs study since the stature does not always seem to be related to thin soil and poor conditions. *Newbould & Jefford* 1680 (Mpanda District, Mahali Mts., Utahya, 20 Aug. 1958) from a damp clay hollow in *Brachystegia* woodland clearly matches the type of *A. archboldiana*. *Greenway & Kanuri* 14511 (Iringa District, Ifuguru [Ifagula], Great Ruaha R, 12 May 1970, is a form with condensed inflorescences and rather larger fruits. Some specimens have very small fruits 1–1.3 mm. wide, e.g. *Tanner* 3132 (Pangani District, Mwera, Langoni, 1 Sept. 1956) or small calyces (Iringa District, Ruaha R. rapids, 20 Oct. 1970, *Richards & Arasululu* 26290; Lushoto District, Korogwe Swamp, 3 Jan. 1963, *Archbold* 119; Lushoto District: Lutindi, July 1893, *Holst* 3441, named *A. baccifera* by Koehne but not cited in E.P.). *Anderson* 1069 (Ulanga District, Ifakara, 8 Aug. 1955) is a form with very condensed inflorescences and fruits 2 mm. wide. The name *A. multiflora* Roxb. (type: India, Calcutta, *Roxburgh* (ubi?)) has been applied to certain East African specimens and Koehne records *A. multiflora* var. *parviflora* (DC.) Koehne from Sudan, Zanzibar and Madagascar and var. *floribunda* (Guill. & Perr.) Koehne from Senegambia to Ethiopia. I have seen nothing that could really be identified with the Indian material unless it is the small-fruited form mentioned above. Without a thorough revision of the Asiatic species I prefer not to join the taxa. What plants Koehne actually referred to his names is not certain, he cites only localities and the material has been destroyed. Although *A. floribunda* Guill. & Perr. was described from Senegambia the name is nowhere mentioned in F.W.T.A.

4. **A. elata** *A. Fernandes* in Bol. Soc. Brot., sér. 2, 48: 125, t. 11 (1974) & in F.Z. 4: 306, t. 76 (1978) & in Fl. Moçamb. 73: 38 (1980). Type: Mozambique, Zambezia, Mathilde, Dumbazi R., *Le Testu* 749 (BM, holo.!)

Aquatic glabrous plant with stout prostrate stem, 4–8 mm. wide (dry), ± 1–1.2 m. long, profusely rooting at the nodes; stem pale, angled; adventitious roots ± 10 cm. long. Leaves narrowly triangular-lanceolate, 2–3.5(–6) cm. long, 0.6–1 cm. wide, narrowly acute at the apex, truncate to slightly rounded auriculate-cordate at the base; lateral venation ± obscure; petiole ± obsolete or scarcely 1 mm. long. Cymes numerous, up to about 1.2 cm. wide in fruit, 7–15(–20 or probably more)-flowered; peduncle 3–5 mm. long; secondary branches ± 2 mm. long; pedicels 1–2 mm. long; bracteoles filiform, minute. Calyx-tube obconic-angular, ± 1(–1.75) mm. long, becoming broadly cupular in fruit, up to 1.5 mm. long, ± 3 mm. wide; lobes 4, broadly triangular, 0.8 mm. tall, ± 2 mm. wide; appendages minute, evident in bud. Petals 4, white (pale lilac in dried material), rounded-obovate, 1.2–1.5 mm. long, 1–1.2 mm. wide, very fugitive. Stamens 4–8. Style and capitate stigma together 0.5 mm. long. Capsule dark red, globose, 2–3 mm. wide, bursting irregularly. Seeds pale brown, concave-convex, 0.5 mm. long. Fig. 12/9–12, p. 39.

KENYA. Teita District: between Kasigau and Maungu, Buguba Rock, 26 Apr. 1963, *Bally* 12701!
DISTR. **K** 7; Mozambique
HAB. In permanent rock-pool; prob. ± 900 m.

NOTE. Possibly only an extreme variant of *A. prieuriana* Guill. & Perr. growing in an extremely favourable environment but the habit is more decumbent and apparently genuinely aquatic, the inflorescences are larger, the leaves more triangular and the petals larger. The measurements in parenthesis refer to the type, still the only other specimen known.

5. **A. linearipetala** *A. Fernandes & Diniz* in Bol. Soc. Brot., sér. 2, 33: 22, t. 3 (1959). Type: Tanzania, Songea District, near Mshangano, R. Luhira, *Milne-Redhead & Taylor* 9239 (K, holo.!, COI, iso.)

Erect branched or unbranched annual herb 10–25 cm. tall; stems reddish, 4-angled below, ± 4-winged above. Leaves oblong or narrowly oblong, 0.8–4.5 cm. long, 0.15–1.3

cm. wide, acute at the apex, auriculate-cordate at the base. Flowers 4-merous, in somewhat lax numerous 5–many-flowered dichasia up to 8 mm. wide in fruit; peduncles 1–3 mm. long; pedicels 0.2–1.5 mm. long; bracteoles minute, up to 1.2 mm. long. Calyx-tube obconic, 0.3–0.5 mm. long; lobes triangular, ± as long as tube; appendages small. Petals white, linear or subulate, ± 0.5 mm. long, ± persistent. Stamens inserted below middle of the tube, ± equalling the lobes. Ovary globose; style plus stigma 0.2–0.3 mm. long. Capsule red, globose, (0.8–)1.2–1.5 mm. diameter. Seeds brown, 0.3 mm. long. Fig. 12/13–17, p. 39.

TANZANIA.   Songea District: N. of Songea, near Mshangano fishponds, waterfall on R. Luhira, 18 Feb. 1956, *Milne-Redhead & Taylor* 9239! & 15 June 1956, *Milne-Redhead & Taylor* 9239A!
DISTR.   **T** 8; not known elsewhere
HAB.   Small pools in rocky river bed; 1030 m.

NOTE.   Possibly close to *A. prieuriana* as stated by Fernandes & Diniz but I suspect it may be a form of *A. senegalensis*; with only one population known its status must remain uncertain.

6. **A. senegalensis** Lam., Tabl. Encycl. 1(2): 311 (1792) & t. 77, fig. 2 (1791); Hiern in F.T.A. 2: 477 (1871), pro parte; Koehne in E.J. 1: 255 (1880) & in E.P. IV. 216: 52, fig. 5D (1903); V.E. 3(2): 647, fig. 282 (1921); F.W.T.A. 1: 144 (1927), pro parte; F.W.T.A., ed. 2, 1: 165 (1954); A. Fernandes & Diniz in Garcia de Orta 6: 100 (1958); Boutique, F.C.B., Lythraceae: 25 (1967); A. Fernandes in C.F.A. 4: 176 (1970); Immelman in Bothalia 21: 40 (1991), pro parte. Type: E. Senegal, *Roussillon* (P, holo., microfiche!)

Erect, prostrate or ascending herb, 4–35 cm. tall; stems 4-angled, occasionally minutely asperulous and sometimes with lowest leaves (? cotyledons) persisting. Leaves oblong, oblanceolate or sublinear, 0.7–5 cm. long, 0.15–1.3 cm. wide, acute or subacute at the apex, ± cordate or cordate-auriculate at the base or sometimes cuneate in basal leaves. Flowers green in 1–many-flowered ± lax cymes; peduncle (0–)1–5 mm. long; pedicels 1–3 mm. long; bracts linear, 2–2.5 mm. long. Calyx campanulate, 1–1.5 mm. long; lobes 4, triangular, 0.3 mm. long; appendages short. Petals 0–4, 0.2 mm. long. Stamens 4. Ovary subglobose, 0.4–0.7 mm. in diameter; style 0–0.3 mm. long. Capsule subglobose, 1–2 mm. in diameter, exceeding the sepals.

var. **senegalensis**

Erect or decumbent annual herb. Leaves ± cordate or cordate-auriculate. Peduncles short to distinct, often exceeding rest of the cyme but occasionally obsolete (as in the type itself). Petals present or absent. Fig. 12/18–21, p. 39.

KENYA.   Tana River District: about 2 km. S. of Ngao, 1 Mar. 1977, *Hooper & Townsend* 1127!
TANZANIA.   Uzaramo District: 28 km. NNW. of Dar es Salaam, 15 July 1972, *Wingfield* 2027! & 13 Aug. 1972, *Wingfield* 2027A! & 6 km. NW. of Dar es Salaam, by Bagamoyo road just beyond Museum Village, 31 July 1971, *Wingfield* 1745!; Zanzibar I., Mwera Swamp, 19 Aug. 1960, *Faulkner* 2696!
DISTR.   **K** 7; **T** 2 (see note), 6; **Z**; Egypt, Senegal, Mali, Sierra Leone, Nigeria, Ethiopia, Zaire, Zambia (fide Immelman), Angola and South Africa
HAB.   Damp ground, seasonal freshwater short grass and sedge swamp on sandy soil, wet sandy silt at edge of drying pools; 0–12(–240) m.

SYN.   *A. filiformis* DC. in Mém. Soc. Phys. His. Nat. Genève 3(2): 95 (1826). Type: Senegal, *Perrottet* (G, holo., BM, iso.!)
    A. floribunda Guill. & Perr., Fl. Seneg. Tent.: 302 (1833). Types: Senegal, marshes around Khann, Cape Verde Peninsula, *Perrottet* 336 (P, syn.!, BM, isosyn.!) & banks of R. Casamance, no collector stated (syn.)*
    A. salsuginosa Guill. & Perr., Fl. Seneg. Tent.: 302 (1833). Type: Senegal, I. Sor, near Saint Louis, ? collector (P, holo.)**
    A. senegalensis Lam. forma *filiformis* (DC.) Hiern in F.T.A. 2: 477 (1871)

NOTE.   *Wingfield* 2027A has 4 small annuals with short styles and one with longer ones up to 1.2 mm. long — it is in advanced fruit but may be *A. auriculata* despite close similarity to the others. He gives petals present on some plants, very pale mauve, up to 1 × 0.8 mm. The restricted distribution of this species is rather strange. It may possibly have been recently introduced. Much older material supposedly of this species has proved to be wrongly determined. The citation for this species is quite wrong in Immelman's revision and elsewhere. It is not mentioned in Encycl. Méth. Bot. 1(1) (1783) which is not by Poiret anyway. Fernandes (C.F.A. 4: 176 (1970)) treats it as published from Poiret, Encycl. Méth. Bot., Suppl. 1: 328 (1810) but Lamarck's description is earlier.

---

* Paris material of Perrottet seen bore no number but is presumably 336; a *Leprieur* specimen with no locality is probably that from Casamance R. (via Herb. Maire, and Herb. Cosson and Herb. Durand). Curiously *A. floribunda* was omitted from F.W.T.A., ed. 2.
** *Perrottet* 334 labelled 'Sénégal, Walo' (BM, P) is presumably not type material.

Vesey-FitzGerald 7069 (Arusha National Park, Maji ya Chai, 23 July 1971, special site hard pan hollow) appears to be a form of *A. senegalensis. Vesey-FitzGerald* 7057 from the same locality is possibly an ephemeral form as was suggested by the collector, but appears very different. It is an erect single-stemmed ephemeral 8 cm. tall with few-flowered axillary cymes, small slightly asperulous leaves with cuneate bases; flowers 4-merous with calyx-appendages evident but apparently no petals; bracteoles oblong; style short. It is certainly not a form of either *A. baccifera* nor *A. auriculata* as it had been named and a certain agreement in the facies, texture of leaves, drying colour suggest 7069 and 7057 are forms of the same species. This only serves to show the extreme difficulty in dealing with collections in this genus since without the collector's suggestion it would have been impossible to name.

var. **ondongana** (*Koehne*) *Verdc.*, comb. nov. Type: Namibia, Amboland, Ondonga, *Rautanen* 206 (Z, holo., H, iso.)

Often prostrate with ascending shoots or ± erect; annual (? always). Leaves cuneate, rounded or cordate at the base. Peduncle usually suppressed but pedicels well developed, the cyme effectively a sessile pseudoumbel. Petals absent.

TANZANIA. Rungwe District: Lake Malawi, Itungi, 26 Sept 1968, *B.J. & S. Harris* 2345!
DISTR. **T** 7; widespread in southern Africa from Mozambique, Malawi and Zambia to Swaziland, Botswana, Angola and Namibia; possibly in W. Africa
HAB. Sandy soil near lake; 450 m.

SYN. *Nesaea ondongana* Koehne in Mém. Herb. Boiss. 10: 78 (1900) & in E.J. 29: 165 (1900) & in E.P. IV. 216: 225 (1903) & in V.E. 3(2): 651 (1921); A. Fernandes & Diniz in Garcia de Orta 6: 104 (1958); Pohnert & Roessler in Prodr. Fl. SW.-Afr. 95: 7 (1966); A. Fernandes in C.F.A. 4: 182 (1970) & in Garcia de Orta, sér. Bot. 2: 78, t. 1 (1975) & in F.Z. 4: 285, t. 69 (1978) & in Fl. Moçamb. 73: 11 (1980); Immelman in Bothalia 21: 41 (1991)
   Ammannia baccifera L. subsp. *intermedia* Koehne in E.J. 41: 79 (1908). Type: Mozambique, Maputo, Matola, *Quintas* 101 (COI, holo.)
   A. intermedia (Koehne) A. Fernandes & Diniz in Garcia de Orta 4: 406 (1956)
   [*A. senegalensis* sensu A. Fernandes & Diniz in Garcia de Orta 4: 406 (1956), pro parte, *non* Lam.]
   A. evansiana A. Fernandes & Diniz in Bol. Soc. Brot., sér. 2, 31: 155, t. 6 (1957). Type: Botswana, Kachikau, Chobe R., *Pole Evans* 4187 (SRGH, holo., IFAN, K, PRE, iso.!)
   Nesaea ondongana Koehne var. *evansiana* (A. Fernandes & Diniz) A. Fernandes in Bol. Soc. Brot., sér. 2, 48: 115 (1974) & in F.Z. 4: 285 (1978); Immelman in Bothalia 21: 42 (1991)
   N. ondongana Koehne var. *beirana* A. Fernandes in Bol. Soc. Brot., sér. 2, 48: 115, t. 1 (1974) & in F.Z. 4: 287, t. 70 (1978). Type: Mozambique, Beira, N. of Macuti Beach, *Noel* 2488 (SRGH, holo., K, LISC, iso.!)
   N. ondongana Koehne var. *orientalis* A. Fernandes in Bol. Soc. Brot., sér. 2, 48: 115, t. 2 (1974) & in F.Z. 4: 287 (1978). Type: Zimbabwe, Chipinga, Chibuwe Pan, *Gibbs Russell* 2083 (COI, holo., K, iso.!)

NOTE. I have been unable to keep *Nesaea ondongana* separate from *Ammannia senegalensis* and although material from southern Africa 'hangs together' very well even some West African material is very similar, e.g. the type material of *A. salsuginosa.* Many specimens from the Flora Zambesiaca area had been previously named *A. senegalensis* by Brenan and others.

7. **A. wormskioldii** *Fisch. & Mey.* in Ind. Sem. Horti Petropol. 7: 42 (1841); Koehne in Fl. Bras. 13(2): 206, t. 40/2 (1877) as *"wormskjoldii"* & in E.J. 1: 256 (1880) & 4: 391 (1883) & in E.P. IV. 216: 53, fig. 5 L (1903) & in V.E. 3(2): 646 (1921); A. Fernandes & Diniz in Garcia de Orta 6: 100 (1958); Pohnert & Roessler in Prodr. Fl. SW.-Afr. 95: 3 (1966); A. Fernandes in C.F.A. 4: 177 (1970) & in F.Z. 4: 308 (1978); Gilbert in Fl. Eth., ined. Type: "Hab. in Brasilia Cap Steudel nomencl. [ed. 2 1: 77 (1841) nomen]" (orig. reference); specimen grown at St. Petersburg from seeds brought from Zaire or Angola (Zaire province), probably collected by Christian Smith (LE, lecto.) [there is a specimen from Angola, *Christian Smith* in BM!]

Annual or rhizomatous perennial herb with erect or ascending unbranched or sparsely branched 4-angled stems 20–60 cm. tall, rooting at the lowest nodes. Leaves narrowly lanceolate to elliptic-obovate, 1–6(–8) cm. long, 0.2–1.7 cm. wide, acute at the apex, truncate, subcordate or subhastate at the base or lowest attenuate-cuneate. Cymes 3–many-flowered, dense, sessile or subsessile; pedicels suppressed or very short; bracteoles linear, 0.4–5 mm. long. Calyx-tube campanulate, 1.5–2 mm. long; lobes 4, very broadly triangular, 0.5–1 mm. long; appendages very short or ± lacking. Petals white, 0.5 mm. long or sometimes lacking (see note). Stamens 4. Ovary globose, 0.7–1.2 mm. in diameter; stigma subsessile or style 0.25–0.3 mm. long at least in fruit. Capsule red, globose, 1.5–2.5 mm. in diameter, exserted from calyx for $\frac{1}{3}$–$\frac{2}{3}$ of its length. Seeds brownish, ± ovate in outline, concave-convex, ± 0.5 mm. long. Fig. 12/22–25, p. 39.

UGANDA. Toro District: Lake George [Ruisamba], east shore, 8 Sept. 1906, *Bagshawe* 1210!; Kigezi District: Ruzhumbura [Ruzumbura], Apr. 1939, *Purseglove* 674!; Mbale District: Sebei, Mbale–Greek R. road, km. 62.4, 17 Jan. 1955, *Norman* 247!

KENYA. Northern Frontier Province: Moyale, 15 July 1952, *Gillett* 13598!; Trans-Nzoia District: SE. Elgon–Saboti [Seboti], Aug. 1967, *Tweedie* 3473!; Laikipia District: 6 km. S. of Suguta Marmar on Rumuruti–Maralal road, 25 Oct. 1978, *Gilbert et al.* 5105!

TANZANIA. Mwanza District: Bwiru, 14 Apr. 1952, *Tanner* 626!; Musoma District: Moru Kopjes, 27 Apr. 1961, *Greenway* 10112! & Ikoma, Mugumu, 7 Apr. 1959, *Tanner* 4105!; Mbeya District: Igawa, 4 Apr. 1962, *Polhill & Paulo* 1990!

DISTR. U 1–4; K 1, 3, 5, 6; T 1, 7; Zaire, Ethiopia, Zambia, Malawi, Zimbabwe, Angola and Namibia

HAB. Marshes on black cotton soil, damp mud, pools in rock crevices and temporary water in granite pavement depressions, often standing in ± 5 cm. of water; 870–1800(–2400) m.

SYN. ?*A. hildebrandtii* Koehne in E.J. 1: 257 (1880) & in E.P. IV. 216, fig. 5K (1903). Type: Kenya, cultivated in Berlin from seeds adhering to the rhizomes of an *Aponogeton** collected in Ukamba by *Hildebrandt* (B, holo.†)
[*A. aegyptiaca* sensu U.K.W.F.: 154 (1974), ? *non* Willd. (see note)]

NOTE. *Hepper & Jaeger* 6683 (Kenya, Laikipia District, Rumuruti, 7 Nov. 1978) is an ephemeral form 5–10 cm. tall in rock pools at 1700 m. There is a major problem which I have left unresolved. Brenan had named most of the material included here as *A. aegyptiaca* Willd. (see note on p. 46) and it is virtually identical with the lowland Egyptian material at Kew including the authentic Delile sheet; *A. aegyptiaca* Willd. lacks petals but several of the East African sheets also lack petals (e.g. *Gilbert et al.* 5627 from K 1, Ndoto Mts., *Tweedie* 1527 from K 3, NE. Elgon and *E.J. & C. Lugard* 536 from Elgon). It is impossible to separate the East African material into taxa having or not having minute petals and I feel certain studies of populations in the field would show both forms in the same population. *A. aegyptiaca* is often completely synonymised with *A. baccifera* or considered a variety or subspecies of it and difficult intermediates do occur. There is no doubt however that the taxon dealt with here is quite different from *A. baccifera*. For this Flora I have, therefore, not succumbed to my original temptation to keep *A. aegyptiaca* as a separate species with *A. wormskioldii* as a variety of it. There is immense scope here for field and laboratory work; they are annuals with plentiful seeds which grow easily. In the meantime I have followed the above course. No authentic material of *A. hildebrandtii* appears to be extant. Although originally describing the style and stigma as equalling the ovary Koehne includes it in his *Astylia* and close to *A. baccifera* which his figure confirms.

8. **A. baccifera** L., Sp. Pl.: 120 (1753) & ed. 2: 175 (1762); Hiern in F.T.A. 2: 478 (1871); C.B.Cl. in Hook. f., Fl. Brit. Ind. 2: 569 (1879); Koehne in E.J. 1: 258 (1880) & 4: 391 (1883) & in E.P. IV. 216: 53, fig. 5M (1903); V.E. 3(2): 647 (1921); Keay in F.W.T.A., ed. 2, 1: 165 (1954); A. Fernandes & Diniz in Garcia de Orta 6: 101 (1958); Pohnert & Roessler in Prodr. Fl. SW.-Afr. 95: 2 (1966); A. Fernandes in C.F.A. 4: 177 (1970); U.K.W.F.: 154 (1974); A. Fernandes in F.Z. 4: 308 (1978) & in Fl. Moçamb. 73: 40 (1980); K. Matthew, Fl. Tamilnadu Carnatic 3, 1: 605 (1983); S.A. Graham in Journ. Arn. Arb. 66: 405 (1985); Kostermans et al., Weeds of rice in Indonesia: 338, fig. 4, 154 (1987); Hewson in Fl. Austr. 18: 97 (1990); Immelman in Bothalia 21: 40 (1991). Lectotype chosen by Graham (1985): China, *Osbeck*; *Linnaean Herb.* 156. 4 (LINN, lecto.) — (see note)

Annual often red erect or ascending simple or ± branched herb or sometimes a minute ephemeral, less often subshrubby and possibly a short-lived perennial, (0.6–)2–80 cm. tall: stems 4-angular or ± winged. Leaves linear to elliptic, 0.7–7 cm. long, 0.15–1(–1.6) cm. wide, narrowed to the apex, nearly always narrowed to a cuneate base, rarely obtuse or subcordate, glabrous or very slightly asperulous. Dichasia (1–)3–many-flowered, ± lax to somewhat dense; peduncles 1(–2) mm. long; pedicels 1–2.5(–4) mm. long; bracteoles minute, linear-lanceolate. Calyx red or green; tube turbinate-campanulate, 1–1.2(–2) mm. long; lobes 4, broadly triangular, 1–1.5 mm. long; intermediary appendages absent or very inconspicuous. Petals nearly always absent or odd minute ones sometimes present. Stamens 4, as long as the calyx-lobes or shorter. Ovary globose, 0.7–1.2 mm. in diameter; style 0.1–0.3 mm. long or stigma sessile. Capsule red, globose, 1–2.5 mm. in diameter, ½–¾-included within the calyx. Seeds brownish, concavo-convex, 0.4 mm. in diameter. Fig. 12/26–30, p. 39.

UGANDA. Karamoja District: Moroto Mt., Mar. 1963, *J. Wilson* 1372!

KENYA. Northern Frontier Province: Ndoto Mts., track up from Ngoronit [Nguronit] Mission, 11 June 1979, *Gilbert et al.* 5629!; Masai District: Mara Game Reserve, Mara Wildlife Research Station, 12 Dec. 1978, *Kuchar* 10211!; Tana River District: Bura, 8 Mar. 1963, *Thairu* 53!

* Presumably *Ouvirandra hildebrandtiana* Eichler, i.e. *Aponogeton abyssinicus* A. Rich.

TANZANIA. Masai District: Ngorongoro Crater, L. Magadi, 5 July 1966, *Greenway & Kanuri* 12534!; Ufipa District: Rukwa N, 15 June 1956, *E.A. Robinson* 1668!; Rufiji, 4 Dec. 1930, *Musk* 16!; Iringa District: near Great Ruaha R., by Great North Road, 17 July 1956, *Milne-Redhead & Taylor* 11238!
DISTR. **U** 1; **K** 1–4, 6, 7; **T** 1–7; throughout tropical Africa from Ethiopia to N. Cape Province, Namibia and Botswana, NE. Africa, Madagascar, Mascarenes, India, Sri Lanka, Java, Philippines, China, Japan and Australia; Europe (introduced), Middle East, Russia, also introduced into the West Indies, etc.
HAB. Silty areas by standing and running water, black soil with grass and sedges by water, seepage zones and pools on rock outcrops in woodland and grassland, also as a weed in rice; 0–1620 m.

SYN. *A. indica* Lam., Tab. Encycl. 1(2): 311 (1792). Type: India, *Sonnerat* (specimen lacking collector's name, P-LAM, ? holo.)
  A. vesicatoria Roxb., Fl. Indica, ed. Carey & Wallich, 1: 447 (1820). Type not stated; *Roxburgh* drawing 35 (K, lecto.!)
  A. attenuata A. Rich., Tent. Fl. Abyss. 1: 278 (1847–8). Types: Ethiopia, R. Djeladjeranne, *Schimper* 778 (P, syn., K, isosyn.!) & R. Mareub, *Quartin Dillon* (P, syn.)

NOTE. Very well-branched specimens with distinctly pedicellate flowers (typical *baccifera*) are very different in appearance from unbranched ± ephemeral specimens with ± sessile flowers in few-flowered inflorescences; these can closely resemble some *Rotala* species but are easily told apart by the sculpture of the capsule wall, e.g. Thika, N. side of Thika R., E. of Nairobi – Fort Hall road, 11 July 1971, *Kabuye* 380. The status of these ephemerals needs investigation; some may be genuine taxa since their habitats and conditions of growth are not always such as would be expected to produce reduced forms. *Verdcourt* 3634 (Kiambu/Machakos Districts, R. Athi at Fourteen Falls, 26 May 1963) has minute petals and might possibly represent *A. hildebrandtii* Koehne, the type of which has been destroyed, although I have associated that name tentatively with *A. wormskioldii. J. Wilson* 1574 (Uganda, Moroto Mt., Nov. 1963, 2400 m.) with rather stiff leaves with subtruncate to ± attenuate leaf-bases matches material from the Sudan and *Lort Phillips* K114 from Somalia (N), Bihen, much of which is named *A. aegyptiaca*. Further material of this high-altitude form is needed. *Faden et al.* 72/153 (Kenya, Tsavo National Park East, Aruba–Buchuma Gate, between signs 149 and 143, 6 Feb. 1972) has distinctly larger capsules than usual, 2.5 mm. wide. Koehne has a complicated very artificial classification of minor taxa.
  Graham (Journ. Arn. Arbor. 66: 405 (1985)) has chosen LINN 156.4 as lectotype; it may possibly be the *Osbeck* specimen Linnaeus cites but is labelled India.
  Koehne, A. Fernandes, K. Matthew and others treat *A. aegyptiaca* Willd. (Hort. Berol. 1: 6, t. 6 (1803) as a subspecies of *A. baccifera*. It is based on a specimen cultivated at Berlin, B-WILLD 3078 (B, syn., microfiche!), presumably from the material grown in Berlin and figured in Hort. Berol.; another specimen labelled 'Aegypto, Delile', possibly the source of the seed (B, syn.), and a sheet at Kew labelled Damiatte, *Delile* (ex. *Herb. Sieber*) is presumably a duplicate. Immelman and others do not separate *A. aegyptiaca* at any level. Brenan, judging by annotations, considered it a distinct species differing in the very condensed inflorescence, leaves not attenuate at the base, larger fruits and more robust habit (see note under *A. wormskioldii*).

9. **A. urceolata** *Hiern* in F.T.A. 2: 478 (1871); Koehne in E.J. 1: 253 (1880) & in E.P. IV. 216: 51, fig. 5 F (1903). Lectotype chosen here (see note): Sudan, Kordofan, Arasch-Cool, *Kotschy* 173 (K, left-hand specimen, lecto.!, BM, GOET, K, isolecto.!)

Erect simple or branched annual 10–30 cm. tall; branches spreading, 4-angled, shortly hispidulous. Leaves linear-oblanceolate, 1.2–6 cm. long, 1.2–9 mm. wide, acute or subacute at the apex, the actual tip ± rounded, gradually narrowed to the base, not auricled, sessile, subscabrid on the margins and single nerve beneath. Cymes dense sessile 1–7-flowered clusters; bracteoles linear, 1.6–2 mm. long, ciliate. Calyx green; tube urceolate, 1.5 mm. long, densely shortly papillate-pubescent; lobes 4(–5), ovate-triangular, 0.7–0.8(–1.2) mm. long and wide, apiculate at apex, widened at base, slightly joined at the extreme base to form a collar 0.6 mm. tall above the constriction of the tube, connivent in fruit, ciliate; appendages minute, forming basal ears to the lobes. Petals not present, Stamens yellow, 4(–5); filaments inserted at unequal heights, 0.5 mm. long, included. Ovary ellipsoid, ± 1 mm. long; style and stigma together 0.2–0.3 mm. long. Capsule subglobose to ellipsoid, 2–3 mm. long, ± 2 mm. wide, enclosed in the calyx-tube which is constricted above it. Seeds 0.35 mm. long. Fig. 12/31–34, p. 39.

KENYA. Tana River District: Tana River National Primate Reserve, Mchelelo, 11 Mar. 1990, *Luke et al.* in *TPR* 124! & same Reserve, Mulondi Swamp, 22 Mar. 1990, *Luke et al.* in *TPR* 780!
DISTR. **K** 7; Somalia (near **K** 1 border) and Sudan (Kordofan)
HAB. Floodplain grassland with *Cordia, Hyphaene, Echinochloa* and edge of swamp with *Terminalia, Combretum, Pluchea, Nymphaea* and sedges; 30 m.

NOTE. Apart from the material cited above and Luke's reference to *Gillett* 2521b from Somalia (not seen) the only other material seen is the original Kotschy material collected in October 1839. It has been essential to select a lectotype since there is some confusion. There are four specimens on Hiern's type sheet at Kew, two with hairy calyx-tubes and two with ± glabrous tubes and I have selected one of the former as lectotype. Hiern shows that he considered all four plants the same species since he states that the calyx is usually covered with numerous small squarrose scales. The sheet bears the original label with misdetermination *Ammannia aegyptiaca* Willd. and locality Arasch-Cool; in front of the printed 173 is inked in '62 and'. Hiern himself cites just 'Kordofan, Kotschy' without numbers or detailed locality. Under *A. urceolata* Koehne cites Arasch-Cool 173 and Abu Gerad 62 in part and under his next species *A. apiculata* gives 62, again in part. I have assumed that the glabrous specimens on the Kew sheet were isosyntypes of this latter species and on a similar mixed sheet at Goettingen Koehne himself has labelled the glabrous specimens as *A. apiculata* mihi floribus valde differt. Hiern of his *A. urceolata* states ovary 1-celled but the Luke material has the ovary 2-celled with dissepiments complete and apparent connection between style and placenta, i.e. technical characters of *Nesaea*; however, there seemed to be no connection in flowers from the lectotype although the dissepiments are probably complete.

DOUBTFUL EPHEMERAL SPECIMENS

Ammannia and *Nesaea* species have, like many semi-aquatics, a propensity for producing ephemeral forms and much more needs to be done in assessing their status. Many are undoubtedly specimens of potentially larger plants growing on poor soil, damp for a short period, but others I suspect are genetically based. This should be born in mind whilst collecting and as much information gathered as possible as to variability in size in one area. Certain specimens I have failed to deal with satisfactorily and they are dealt with below in the hope that they can be recollected and field studies made.

Nyakundi 273 (Kenya, Masai District, Ol Doinyo Orok, 22 Mar. 1986, at 1750–2000 m, rock crevice in river), an erect unbranched ephemeral 9 cm. tall has the leaves subtruncate at the base. Cymes at first ± sessile, 1–2-flowered, pedicels up to 2 mm. long in fruiting state. Calyx-appendages evident, 0.3 mm. long. Petals probably absent. Style 0.2 mm. long. Originally named *A. baccifera* and known from only the single specimen, more material is essential to establish its identity.

Pedersen 1022 (Iringa, just E. of College of National Education, 24 July 1972, 1600 m., common and dominant in and near a temporary waterhole), an ephemeral 8–17 cm. tall, has the ± cuneate leaf-bases of *A. baccifera* but the style is too long and 4 pale purple petals 1 × 1 mm. are present in at least some flowers. Perhaps best placed as a form of *A. prieuriana* with atypical leaf-base.

Gillett 20149 (Kenya, Meru National Park, Kiolu R. crossing, 25 Dec. 1972) had been named *A. baccifera* but quite large petals present, style rather longer and leaves subtruncate at the base — perhaps an ephemeral form of *A. prieuriana* but style too short for the typical '*archboldiana*' form; it is too long for *A. hildebrandtii* e descr.

8. HIONANTHERA

A. Fernandes & Diniz in Bol. Soc. Brot., sér. 2, 29: 90 (1955)

Annual or perennial aquatic or terrestrial glabrous herbs. Leaves decussate, sessile, linear, usually widened at base or sometimes basal pair narrowly oblong; midrib impressed above; basal pair of narrowly oblong leaves present in one species. Flowers (3–)4(–5)-merous, sessile in dense axillary clusters partially enclosed by the widened leaf-bases; outer bracteoles 2, narrowly lanceolate, hyaline, ± as long as young clusters, stipule-like; inner bracteoles whitish, numerous, subulate. Calyx campanulate, scarious, 8-nerved below, 4-nerved above the insertion of the stamens; lobes very broadly triangular; sinus-appendages short. Petals persistent, corrugated. Stamens (3–)4(–5), opposite the lobes; anthers and pollen violet. Ovary sessile or shortly stipitate; dissepiments interrupted above the column of the placenta which is not continuous with the style, incompletely 2-locular, 2–5-ovulate; style slender, much longer than ovary. Capsule thinly membranous, dehiscing irregularly; wall not densely striate. Seeds dark violet, few, concavo-convex.

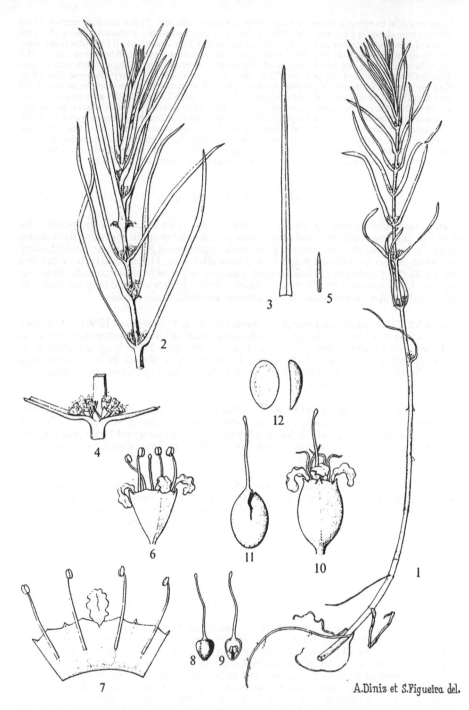

A.Diniz et S.Figueira del.

FIG. 14. *HIONANTHERA MOSSAMBICENSIS*—**1**, habit, × ½; **2**, upper part of stem, × 1; **3**, leaf, × 1; **4**, part of stem showing floral glomerules; **5**, outer bracteole, × 4; **6**, lateral view of flower, × 10; **7**, calyx, opened to show appendages of calyx, one petal, stamens and nectariferous ring, × 10; **8**, pistil, × 10; **9**, longitudinal section of ovary, × 10; **10**, fructiferous calyx, × 10; **11**, capsule dehiscing, × 10; **12**, seeds, × 10. All from *Torre* 719. Drawn by Diniz & Figueira. Reproduced with permission from Flora Zambesiaca.

Fernandes & Diniz described four very similar 'species' but I think only one or perhaps two should be recognised, occurring in S. Tanzania, Mozambique and Zimbabwe. The genus was previously unrecorded for East Africa.

H. mossambicensis A. *Fernandes & Diniz* in Bol. Soc. Brot., sér. 2, 29: 91, t. 6 (1955) & in Garcia de Orta 4: 398, t. 4 (1956); A. Fernandes in F.Z. 4: 309, t. 77 (1978) & in Fl. Moçamb. 73: 41 (1980). Type: Mozambique, Nampula, *Torre* 719 (COI, holo., LISC, iso., K, photo.!)

Annual or perennial branched or unbranched erect or decumbent herb, 12–45 cm. tall, often growing in water; stems sometimes caespitose. Leaves linear, 2.5–6 cm. long, 1.5–3 mm. wide, acute at the apex, usually widened at the base to up to 3 times the width at the middle but some leaves scarcely if at all widened; midrib impressed or channelled above, prominent beneath and often margined. Nodal pair of flower clusters up to 8 mm. across; pedicels obsolete or up to 1 mm. long (*fide* Fernandes); large bracteoles 3–3.5 mm. long, 0.7–1.5 mm. wide, inner ones ± 1 mm. long. Calyx campanulate; tube 1.5 mm. long, 1 mm. wide; lobes and appendages very short. Petals lilac, violet or reddish, oblong, 1–1.25 mm. long. Stamens ± 4, 2–2.3 mm. long, inserted near the middle of the tube and exserted ± 1–1.5 mm. long; style 1.5–2 mm. long, almost as long as the stamens. Capsule ellipsoid, ± 2 mm. long, scarcely exserted. Seeds dark violet, 2–4, 1–1.75 mm. long, 0.5–1 mm. wide. Fig. 14.

TANZANIA Dodoma District: about 37 km. on Itigi–Chunya road, 25 Mar. 1965, *Richards* 19864!
DISTR. T 5; Mozambique and Zimbabwe
HAB. Small pools on flat granite rocks in wet grassland; 1500 m.

SYN. *H. graminea* A. Fernandes & Diniz in Bol. Soc. Brot., sér. 2, 29: 92, t. 7 (1955) & in Garcia de Orta 4: 398, t. 5 (1956); A. Fernandes in F.Z. 4: 311 (1978). Type: Mozambique, Nampula, *Torre* 715 (COI, holo., LISC, iso., K, photo.!)
H. torrei A. Fernandes & Diniz in Bol. Soc. Brot., sér. 2, 29: 93, t. 8 (1955) & in Garcia de Orta 4: 398, t. 6 (1956); A. Fernandes in F.Z. 4: 311 (1978). Type: Mozambique, Nampula, *Torre* 1212 (COI, holo., LISC, iso., K, photo.!)

NOTE. I have not included *H. garciae* A. Fernandes & Diniz (in Bol. Soc. Brot., sér. 2, 29: 94, t. 9 (1955) & in Garcia de Orta 4: 398, t. 7 (1956); A. Fernandes in F.Z. 4: 311 (1978); type: Mozambique, Serra de Bandula, near Chimoio, *Garcia* 790 (LISC, holo., COI, K!, MO, P, PRE, SRGH, iso.)) in the above synonymy since most specimens have a distinctive pair of narrowly oblong basal leaves up to 4 mm. wide although they are absent in *Biegel* 1885 (Zimbabwe, Gwelo), a specimen annotated by Fernandes. The Richards specimen cited above shows a number of galled flowers drying dark and ± 3 mm. long and similar galls are apparent on *Chiparawasha* 413 (Zimbabwe, Masvingu District, Gt. Zimbabwe). The leaves of the Richards specimen are seen to be closely micropunctate under high magnifications and specimens from elsewhere show similar but less evident punctures. Fernandes gives the dimension of the capsule of *H. graminea* as 0.5 mm. long but this must be wrong since the seeds are thus bigger than the capsule.

9. ROTALA*

L., Mant. Pl. Alt.: 143, 175 (1771); C.D.K. Cook in Boissiera 29: 1–156 (1979)

Aquatic, amphibious or terrestrial annual or perennial glabrous herbs with creeping to erect or floating stems. Leaves opposite, whorled or rarely alternate, simple, entire, sessile or rarely shortly petiolate; stipules absent. Flowers regular, hermaphrodite, solitary in leaf-axils (bracts) along the main axis or in lateral or terminal racemes, sometimes heterostylous, occasionally cleistogamous; bracteoles 2 (rarely more in one non-African species) or absent. Calyx tubular, often enclosing the ovary; lobes 3–6, valvate, persistent; appendages or folds sometimes present between the lobes. Petals up to 6, minute to large and showy or sometimes absent, inserted at top of calyx-tube, usually crumpled in bud, entire, erose or pinnately divided in one species. Stamens 1–6, inserted in lower half of calyx-tube, sometimes basal and appearing free, occasionally replaced by staminodes. Ovary superior, 2–4-locular; placentation axile, becoming free-central at maturity; style simple, persistent on one fruit valve; stigma capitate (bilobed in one Indian species). Capsule septicidally dehiscent by 2–4 valves; valves with microscopic horizontal striations. Seeds few–numerous, ± ellipsoidal, with mucilaginous hairs.

---

* Description of genus taken entirely from Cook's revision and rest of account very largely based on it.

Cook recognises 44 species and two further are uncertain; for the world-wide revision he has sunk several of Fernandes and Diniz's species proposed on more parochial grounds. Although needing high powers to see, the horizontal striation on the capsule valves is an excellent diagnostic character.

1. Flowers in terminal raceme-like inflorescences clearly separate from the densely leafy part of the stem . . . . . 2. *R. repens*
   Flowers axillary along the leafy stems 2
2. Flowers up to 20 in axillary clusters at each node; usually strictly erect annual with linear-elliptic petiolate leaves; bracteoles absent . . . . 6. *R. serpiculoides*
   Flowers solitary in axils 3
3. Appendages present between the calyx-lobes; leaves lanceolate 4
   No appendages between the calyx-lobes 5
4. Sepals 5; petals absent (**T** 4) . . . . 4. *R. verdcourtii*
   Sepals and petals 3 (**T** 2) . . . . 5. *R. juniperina* var.
5. Petals absent; leaves in whorls of 3–8 . . . . 1. *R. mexicana*
   Petals present or absent; leaves opposite, decussate or rarely in whorls of 3 at stem apex 6
6. Capsule opening by 4 valves; fruiting calyx subglobose, ± 1.5 mm. long . . . . 3. *R. tenella*
   Capsule opening by 2 or 3 valves 7
7. Capsule opening by 3 valves; petals absent 8
   Capsule opening by 2 valves 9
8. Calyx-lobes 3–5; calyx-tube 0.5–0.75 mm., with lobes 0.25–0.5 mm. long; leaves mostly linear . . . . . . 1. *R. mexicana*
   Calyx-lobes 4 (rarely 5); calyx-tube ± 1 mm. long, with lobes about the same length; leaves mostly elliptic to round . . 7. *R. gossweileri*
9. Stamens (1–)2(–3); style 0.5 mm. long or less 10
   Stamens 4 (rarely less); capsule longer than calyx; style exceeding 0.5 mm. where known 11
10. Fruiting calyx-tube at least 1–1.5 mm. long; capsule less than twice as long as calyx-tube; style 0.25–0.5 mm. long . . . . 8. *R. filiformis*
    Fruiting calyx-tube 0.3–0.5 mm. long; capsules about 4 times as long as the calyx-tube; style 0.1–0.2 mm. long . . . . 9. *R. capensis*
11. Petals usually 4; mature capsule ellipsoid, longer than total length of calyx, 2 mm. long or more, much exceeding the lobes; seeds 0.75–1 mm. long; annual or perennial herb with branched stems to 25 cm. . . . . . 10. *R. lucalensis*
    Petals absent; capsule described as globose-ellipsoid "semi-exserted beyond the calyx-lobes"; gregarious small annual with simple stems to 7 cm. (known only from description) . . . . 11. *R. stuhlmannii*

1. **R. mexicana** *Cham. & Schlecht.* in Linnaea 5: 567 (1830); Koehne in Fl. Bras. 13: 195, t. 39, fig. IIa, b (1877) & in E.J. 1: 150 (1880) & in E. & P. Pf. 3(7): 7, fig. 2A–E (1891) & in E.P. IV. 216: 29, fig. 12 (1903); F.P.S. 1: 142 (1950); F.W.T.A., ed. 2, 1: 164 (1954); C.D.K. Cook in Boissiera 29: 33, fig. 4A–O, map 3 (1979) (full extra-African synonomy). Type: Mexico, near Hacienda de la Laguna, *Schiede & Deppe* 566 (HAL, holo., LE, MO, iso.)

Aquatic, amphibious or terrestrial perennial herb with often reddish simple or branched creeping ascending erect or floating stems 1–15 cm. long, rooting at nodes. Leaves decussate or in whorls of 3–8; submerged leaves linear, up to 1.5 cm. long, ± 0.5 mm. wide; aerial leaves linear to narrowly ovate, 0.2–1.5 cm. long, 0.2–2.5 mm. wide. Flowers reddish or whitish, sessile and solitary in the axils, occasionally cleistogamous. Calyx-tube golden red or pink, ± subglobose, 0.5–0.75 mm. long; lobes (3–)4(–6), triangular, 0.25–0.5 mm. long; appendages absent. Petals absent. Stamens 1–4, usually included within the calyx-tube, occasionally exserted. Ovary globose; style usually very short, rarely up to 0.3 mm.; stigma capitate. Capsule globose red to purple, occasionally spotted white, 0.5–0.75 mm. in diameter, opening by 3 valves. Seeds semiovoid, ± 0.3 mm. long. Fig. 15/1–5, p. 53.

UGANDA. Teso District: Serere, Dec. 1931, *Chandler* 319!
KENYA. Uasin Gishu District: Kaptagat, Jan 1962, *Tallantire* 602!
TANZANIA. Tanga District: Lwengera R., 6.4 km. ENE. of Korogwe, 27 June 1953, *Drummond & Hemsley* 3044!; Chunya District: Lupa N. Forest Reserve, 3 June 1963, *Boaler* 974!; Songea District: valley near R. Mtanda about 9.5 km. SW. of Songea, 24 June 1956, *Milne-Redhead & Taylor* 10890! & just below Lumecha Bridge, 4 May 1956, *Milne-Redhead & Taylor* 9387A!
DISTR. U 3; K 3; T 2–4, 7, 8; from Senegal to Sudan and Ethiopia, Transvaal and Namibia and virtually throughout the warmer parts of the world but lacking from Congo Basin, Arabia and Pacific Is.
HAB. Seasonally dry swampy areas, riverine flats, on bare soil between grasses, also as an aquatic in water up to 15 cm. deep on seasonally flooded boggy ground; 300–1400 m.

SYN. *R. pusilla* Tul. in Ann. Sci. Nat., sér. 4, 6: 128 (1856); H. Perrier in Fl. Madag. 147, Lythracées: 11 (1954); A. Fernandes & Diniz in Garcia de Orta 6: 92 (1958); A. Fernandes in C.F.A. 4: 168 (1970) & in F.Z. 4: 322 (1978). Types: Madagascar, near Diego-Suarez, *Bernier* 383 & *Boivin* 2692 bis (P, syn.)
   [*R. verticillaris* sensu Hiern in F.T.A. 2: 467 (1871), *non* L.]
   R. mexicana Cham. & Schlecht. subsp. *hierniana* Koehne in E.J. 1: 151 (1880). Type: Sudan, Seriba Ghattas, *Schweinfurth* 2434 (B, syn., K, isosyn.!)
   R. mexicana Cham. & Schlecht. subsp. *pusilla* (Tul.) Koehne in E.P. IV. 216: 30 (1903)

NOTE. Koehne cites several specimens under his subsp. *hierniana* including "? Mad. sec. Tul. (sub *R. pusilla*)" which presumably is invalidated as a syntype due to the '?' — if the subspecies is kept up it presumably should be called subsp. *hierniana* rather than subsp. *pusilla*. Cook's citation of *Boivin* 2692 as holotype of *R. pusilla* can perhaps be taken as a lectotypification but I am not sure since Tulasne distinctly gives 2692 bis. There is at least one 6-lobed calyx on *Boaler* 974, but Cook gives only 3–5 in his description.

2. **R. repens** (*Hochst.*) *Koehne* in Verh. Bot. Ver. Brandenb. 22 (Sitz. 73): 24 (1880); Boutique, F.C.B., Lythraceae: 4 (1967); U.K.W.F.: 155 (1974); C.D.K. Cook in Boissiera 29: 45, fig. 7A–C (1979); M.G. Gilbert in Fl. Eth., ined. Type: Ethiopia, Tigre, Sana, Docheli, *Schimper* 729 (B, holo., BM!, BR, G, K!, L, LE, M, MEL, P, Z, ZT, iso.))

Aquatic perennial of podostemaceous appearance with branched reddish stems 5–50 cm. long; roots attached to rocks. Leaves often reddish, submerged, dense and numerous, alternate, opposite or whorled on the same stem, capillary, 0.8–3.5 cm. long, 0.1–0.5 mm. wide, acute at apex. Flowers red or purplish, borne in emergent terminal racemes 2–6 cm. long; peduncle 0.5–2.5 cm. long; pedicels 1–5 mm. long; bracts linear to lanceolate, 1.3–1.8 mm. long; bracteoles 2, lanceolate, 0.8–1.5 mm. long. Calyx-tube campanulate, 1 mm. long; lobes 4, broadly triangular, 0.1–0.25 mm. long; appendages and folds absent. Petals white or red, 4, rhombic, 0.1–0.2 mm. long, not or scarcely exceeding the calyx-lobes. Stamens 4, crimson, purple or orange, the filaments 1.5 mm. long, well-exserted. Ovary ellipsoid to ovoid, 0.5–0.6 mm. long; style 0.4–1 mm. long; stigma obconic. Capsule ovoid, ± 1.5 mm. long, opening by 2 valves. Seeds flattened ellipsoid, 0.5 mm. long. Fig. 15/6–10, p. 53.

UGANDA. Karamoja District: Moroto Mt., SE. side, Nov. 1963, *J. Wilson* 1587!; Elgon, Sipi Falls, Dec. 1930, *J.R.M. Wallace* 10a! & Bugisu, Greek R. Camp, Jan. 1936, *Eggeling* 2496! & Bugisu, Kaburoron [Kaburon] Jan. 1948, *Eggeling* 5729!
UGANDA/KENYA. Mbale/Trans-Nzoia Districts: Suam R., Karamoja Drift, 18 Feb. 1935, *G. Taylor* 3412!
DISTR. U 1, 3; ?K 3; Ethiopia and Zaire (Shaba) — see note
HAB. Attached to rocks in running water, sometimes in the spray of waterfalls; 1200–1800 m.

SYN. *Rhyacophila repens* Hochst. in Flora 24: 659 (1841); Hiern in F.T.A. 2: 470 (1871)
   Quartinia turfosa A.Rich., Tent. Fl. Abyss. 1: 277 (1847–8), t. 51 (1851), *nom. illegit.* Type: as for *Rotala repens*

NOTE. Boutique records this species from Shaba based on *Quarré* 5564. Cook does not mention this specimen but cites one, the label of which he could not read, from 'Kurudelungu' obviously Kundelungu in Shaba but puts the specimen under Ethiopia. Since the extension of distribution by 1450 km. to S. Zaire is surprising I have investigated the matter. This specimen is of course correctly named and a mix-up of labels appears impossible - a genuine discontinuity is the only explanation. Quarré's label records the specimen was collected in the R. Lukafu on the mountain.

3. **R. tenella** (*Guill. & Perr.*) *Hiern* in F.T.A. 2: 467 (1871); Koehne in E.J. 1: 170 (1880) & in E.P. IV. 216: 39 (1903); V.E. 3 (2): 645 (1921); F.P.S. 1: 141 (1950); F.W.T.A., ed. 2, 1: 164 (1954); A. Fernandes & Diniz in Garcia de Orta 3: 197 (1955); F.W.T.A., ed. 2, 1: 760 (1958);

Boutique, F.C.B., Lythraceae: 5 (1967); U.K.W.F.: 155 (1974); C.D.K. Cook in Boissiera 29: 63, fig. 12 A–K (1979); Maquet in Fl. Rwanda 2: 484 (1983). Type: Senegal "Walo, prope Richard-Tol, la Sénégalaise etc." *Perrottet 333* (P, lecto., BM, G, LE, isolecto.)*

Aquatic or amphibious annual or ? perennial, sometimes entirely submerged; stems pink, inflated and ± fleshy when submerged, floating or creeping, usually erect and scarcely branched above, rooting at the nodes, 5–40 cm. long. Leaves decussate, sessile; submerged ones elliptic to oblong or obovate, 0.5–2 cm. long, 2–7 mm. wide, obtuse at the apex, cordate at the base, membranous; aerial ones oblong to obovate, 0.3–1.2 cm. long, 2–6 mm. wide, obtuse at the apex, cordate at the base. Flowers pink, reddish or greenish, solitary, axillary; pedicels 0.3–2 mm. long; bracteoles 2, filiform to linear-lanceolate, 0.25–1 mm. long. Calyx-tube ± cylindrical to urceolate, 0.8–1.2 mm. long, enlarging and becoming subglobose and often splitting in fruit; lobes 4, triangular, 0.5–0.9 mm. long, ± mucronate; interjected folds distinct before fruiting. Petals pink or white, 0–4, sometimes rudimentary, linear to obovate, 0.2–0.7 mm. long. Stamens 4 or sometimes less. Ovary ellipsoid, 1 mm. long, 4-locular; style very short, 0.2–0.3 mm. long, stigma capitate. Capsule ellipsoid to globose, (1.5–)2–3 mm. long, equalling or exceeding the calyx, 4-valved. Seeds compressed-ovoid, 0.35–0.6 mm. long. Fig. 15/11–14.

UGANDA. Karamoja District: Dodoth, Napori [Napore] Hills, Karenga, Nov. 1972, *J. Wilson* 2168!; Teso District: Serere, Dec. 1931, *Chandler* 320!; Mengo District: Kampala, Ngulu Hill, Aug.1937, *Hancock & Chandler* 1843!
KENYA. Uasin Gishu District: 13 km. Eldoret–Kitale, 8 Oct. 1981, *Gilbert & Tadessa* 6502!; S. Nyeri District: Mwea-Tebere Scheme, Nguka village, 14 Nov. 1966, *D. Wood* 751!; Nairobi National Park, near Impala Point, 21 Jan. 1962, *Verdcourt & E.J. Brown* 3237!
TANZANIA. Mwanza District: ? Burru, 30 Mar. 1952, *Tanner* 619!; Moshi District: Mpololo, Aug. 1928, *Haarer* 1528!; Dodoma District: Kazikazi, 10 June 1932, *B.D.Burtt* 3705!; Mbeya District: Great North Road, Igawa, 13 Apr. 1962, *Polhill & Paulo* 2004!
DISTR. U 1–4; K 3, 4; T 1, 2, 5, 7; Senegal to Nigeria, Gabon, Sudan S. to South Africa (Transvaal), Botswana and Namibia, also in Madagascar
HAB. Seasonal pools and seasonally dry swamps, particularly shallow pools on thin soil on rock outcrops in dry scrub and wooded grassland ('wet pans'); in up to 30 cm. of water; also waterholes and a serious weed in young rice; 810–1700 (–2015) m.

SYN. *Ammannia tenella* Guill. & Perr., Tent. Fl. Seneg.: 297 (1833)
    Rotala brevistyla Bak.f. in J.L.S. 37: 153 (1905); Koehne in E.J. 41: 77 (1907); A. Fernandes & Diniz in Bol. Soc. Brot., sér. 2, 30: 106, t. 1 (1956). Type: Uganda, Ankole District, Mulema, *Bagshawe* 316 (BM, holo.!)
    R. oblonga A. Peter in Abh. Ges. Wiss. Göttingen N.F. 13: 87, fig. 18 (1928). Type: Tanzania, Dodoma District, L. Chaya, *Peter* 34143A (B, holo. & iso.)
    R. submersa Pohnert in Mitt. Bot. Staats. München 1: 449 (1954); Pohnert & Roessler in Prodr. Fl. SW.-Afr. 95: 9 (1966); A. Fernandes in Bol. Soc. Brot., sér. 2, 48: 128 (1974) & in F.Z. 4: 316 (1978). Type: Namibia, Amboland, Kanovley, *Dinter* 7281 (M, holo., B, iso.)
    R. tenella (Guill. & Perr.) Hiern forma *fluviatilis* A. Fernandes & Diniz in Garcia de Orta 3: 197 (1955). Type: Guinea Bissau, Boé, Dandum, *Espírito Santo* 2370A (COI, holo.)
    R. peregrina H. Perrier** in Not. Syst., Paris 14: 308 (1953) & Fl. Madag. 147, Lythracées: 12 (1954). Types: Madagascar, Antisirabe, *Perrier* 6596 & 18476 (P, syn.)
    R. pedicellata A. Fernandes & Diniz in Bol. Soc. Brot., sér. 2, 31: 152, t. 2 (1957) & 48: 128 (1974). Type: Zambia, Mapanza, *E.A. Robinson* 622 (SRGH, holo., K, iso.!)
    R. submersa Pohnert var. *angustipetala* A. Fernandes in Bol. Soc. Brot., sér. 2, 48: 128, t. 14 (1974) & in F.Z. 4: 317 (1978). Type: Zambia, Mbala, *Michelmore* 434 a (K, holo.!)
NOTE. I have followed Cook's broad interpretation of this species.

* From Guillemin & Perrottet's original citation it is clear more than one locality is involved; I have accepted Cook's citation of *Perrottet 333* as 'holotype' as a lectotypification.
** Cook claims this is an invalid name but I do not see why; he also states some of the cited specimens (Perrier cites 7 in all) are not *Rotala*. I have therefore cited only the two which Cook lists as this species.

FIG. 15. *ROTALA MEXICANA*—1, habit, ×⅔; 2, leaf, ×3; 3, flower, ×12; 4, corolla, opened out, × 12; 5, ovary, × 12. *R. REPENS*—6, habit, × ⅔; 7, leaf, × 3; 8, flower, × 12; 9, corolla opened out, ×12; 10, ovary, ×12. *R. TENELLA*—11, leaf, ×3; 12, flower, ×12; 13, corolla, opened out, × 12; 14, ovary, × 12. *R. VERDCOURTII*—15, leaf, × 3; 16, flower, × 12; 17, corolla, × 12; 18, ovary, × 12. *R. JUNIPERINA*—19, leaf, × 3; 20, flower, × 12; 21, corolla, opened out, × 12; 22, ovary, × 12. *R. SERPICULOIDES*—23, leaf, × 2; 24, inflorescence, × 4; 25, flower, × 12; 26, corolla, opened out, × 12; 27, ovary, × 12. *R. GOSSWEILERI*—28, leaf, ×3; 29, flower, ×12; 30, corolla, opened out, ×12; 31, ovary, ×12.

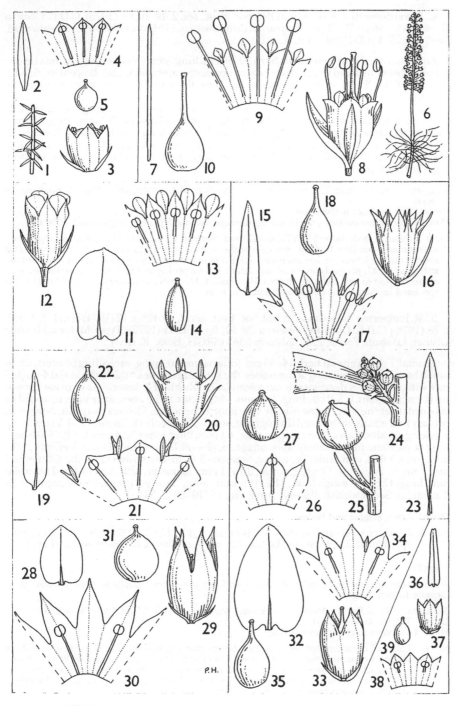

R. FILIFORMIS—**32**, leaf, × 3; **33**, flower, × 12; **34**, corolla, opened out, × 12; **35**, ovary, × 12. *R. CAPENSIS*—**36**, leaf, × 3; **37**, flower, × 12; **38**, corolla, opened out, × 12; **39**, ovary, × 12. 1–5, from *Drummond & Hemsley* 3044; 6–10, from *Eggeling* 2496; 11–14, from *B.D. Burtt* 3705; 15–18, from *Verdcourt* 2854; 19–22, from *Bigger* 2184; 23–27, from *Greenway & Kanuri* 14457; 28–31, from *Chandler & Hancock* 30; 32–35, from *Milne-Redhead & Taylor* 10815; 36–39, after Clarke. Drawn by Pat Halliday.

4. **R. verdcourtii** A. *Fernandes* in Bol. Soc. Brot., sér. 2, 49: 10, t. 3 (1975); C.D.K. Cook in Boissiera 29: 90 (1979). Type: Tanzania, Buha District, 115 km. on Kibondo–Kasulu road, *Verdcourt* 2854 (COI, holo., EA, K, iso.!)

Aquatic annual (? or perennial) herb 20–30 cm. long; stems submerged, leafless below, usually ± branched and leafy above. Leaves opposite (rarely alternate), lanceolate, 0.3–1.5 cm. long, 0.5–1.5 mm. wide, attenuate at the apex, the actual tip truncate or emarginate, attenuate at the base, spreading or reflexed, 1-nerved. Flowers pink, 5-merous, sessile, solitary in the axils; bracteoles linear-filiform, 1.25 mm. long, exceeding the calyx-tube. Calyx campanulate-urceolate; tube ± 1 mm. long; lobes triangular, 0.4 mm. long, 0.5 mm. wide, acute; appendages narrowly triangular, 0.3 mm. long. Petals absent. Stamens 2; filaments ± 0.4 mm. long. Ovary ellipsoid, 0.5 mm. wide, 3-locular; style 0.25 mm. long. Capsule reddish, subglobose, 1–1.2 mm. long. Seeds yellowish, ellipsoid, 0.25 mm. long. Fig. 15/15–18, p. 53.

TANZANIA.   Buha District: 115 km. on Kibondo–Kasulu road, near Mugombe, 15 July 1960, *Verdcourt* 2854!
DISTR.   **T** 4; not known elsewhere
HAB.   Clear stream on fine gravel running through swampy *Hyparrhenia* grassland; 1350 m.

NOTE.   *Braun* in *Herb. Amani* 5438 (Tabora District, Nyembe, 6 June 1913, wet meadows), a specimen unfortunately not seen by Cook, has exactly the floral structure of *R. verdcourtii* and keys to that species but the leaves are less attenuate, 1.5–3.5 mm. wide, with more evident venation and the style shorter; it is probably only a minor variant. It is surprising that no further material has been obtained from so well-collected an area. Mr. Masinde kindly drew my attention to the fact that this specimen keyed to *R. verdcourtii* and sent it on loan.

5. **R. juniperina** A. *Fernandes* in Bol. Soc. Brot., sér. 2, 48: 126, t. 12 (1974) & in F.Z. 4: 314, t. 78 (1978); C.D.K. Cook in Boissiera 29: 97, fig. 19E–H (1979). Type: Malawi, Mulanje District, Likabula Forest, *E.A. Robinson* 5353 (SRGH, holo., K, iso.!)

Annual, ? or perennial herb 14–30 cm. long; stems creeping, ascending or erect, often much branched and rooting at the nodes, the submerged ones ± thick, the aerial 4-angled or ± winged. Leaves decussate or occasionally in whorls of 3, lanceolate to ovate, largest on the main stems, 0.3–2 cm. long, 1–8 mm. wide, acute at the apex, attenuate, rounded or subcordate at the base, subsessile or very shortly petiolate. Flowers purplish, 3-merous, solitary in the axils, distributed throughout, sessile; bracteoles 2, linear, 1.2–1.5 mm. long, ± as long as calyx. Calyx-tube campanulate, ± 1 mm. long; lobes 3, triangular, 0.35–0.5 mm. long (± papillate in Tanzania); appendages 3, (0.4–)0.75 mm. long. Petals 3, elliptic, ± 0.5 mm. long, ± 0.25 mm. wide, persistent. Stamens 3 (2 in Tanzanian var.), the filaments 1 mm. long in Tanzania. Ovary globose, 0.75–1 mm. diameter; style 0.25 (0.4 in Tanzania) mm. long. Capsule subglobose, (1.2–)1.5 mm. long, 3-valved (the valves ± papillate in Tanzania). Seeds whitish, 0.5 mm. long. Fig. 15/19–22, p. 53.

DISTR.   Zaire, Zambia and Malawi

SYN.   *R. decumbens* A. Fernandes in Bol. Soc. Brot., sér. 2, 48: 127, t. 13 (1974) & in F.Z. 4: 314 (1978). Type: Zambia, Kambulamwanda dam, 128 km. N. of Choma, *E.A. Robinson* 723 (K, holo.!)

var.?

Plant very sparsely branched. Calyx-lobes and capsule-valves ± papillate. Style 0.4 mm. long. Capsule 1.2 mm. long.

TANZANIA.   Moshi District: Moshi–Arusha road, Masama turn-off, 9 Sept. 1968, *Bigger* 2184!
DISTR.   **T** 2; not known elsewhere
HAB.   In shallow stagnant water; 990 m.

NOTE.   This had been equated with my 2854 before that was described as *R. verdcourtii* and the resemblance is certainly strong but it seems more closely related to *R. juniperina*. With only one specimen of each to examine no firm conclusions can be drawn as to their correct relationship. Water plants vary so much according to differences in habitat that it is particularly difficult to decide taxonomic status from inadequate material.

6. **R. serpiculoides** *Hiern* in F.T.A. 2: 469 (1871); Koehne in E.J. 1: 158 (1880); Hiern in Cat. Afr. Pl. Welw. 1: 371 (1898); Koehne in E.P. IV. 216: 33, fig. 1M (1903); V.E. 3(2): 643 (1921); F.P.S. 1: 142 (1950); A. Fernandes & Diniz in Garcia de Orta 6: 92 (1958), Boutique, F.C.B., Lythraceae: 4 (1967); A. Raynal in Adansonia, N.S. 7: 544 (1967); A. Fernandes in C.F.A. 4: 168 (1970) & in F.Z. 4: 314 (1978); U.K.W.F.: 155 (1974); C.D.K. Cook in Boissiera

29: 101 (1979); Vollesen in Opera Bot. 59: 52 (1980); Maquet in Fl. Rwanda, 2: 484, fig. 152/2 (1983). Type: Angola, Lubango, near Monhino, *Welwitsch* 2355 (LISU, holo., BM, iso.)

Erect or decumbent simple or branched annual, 3–30 cm. tall; stem green or tinged with red. Leaves narrowly elliptic-oblong to oblanceolate, decussate, 0.5–2.5 cm. long, 1–6 mm. wide, obtuse at the apex, narrowed at the base into a short petiole, spreading or suberect. Flowers crimson or reddish purple, 3–4-merous, (1–)3–12 in axillary sessile racemose clusters; rarely with vegetative growth continued beyond the inflorescence rhachis; pedicels 0.4–0.5(–1.3; –2 *fide* Cook) mm. long; bracteoles linear-subulate ± 1 mm. long. Calyx-tube usually campanulate, ± 0.6 mm. long; lobes 4(–5)*, triangular, 0.4–0.7 mm. long, acuminate; appendages absent. Petals absent, (or 1, white, *fide Milne-Redhead & Taylor* 10873). Stamens 1(–3), yellow, included. Ovary reddish, ellipsoid, 0.5–1 mm. long; style very short; stigma cream, capitate. Capsule ellipsoid or globose, 1–1.5 mm. long, 3(–4)*-valved. Seeds brownish, subglobose or compressed-ellipsoidal, concavo-convex, 0.25–0.5 mm. long. Fig. 15/23–27, p. 53.

UGANDA. Toro District: Queen Elizabeth National Park, Kasenyi, 16 Dec. 1967, *Lock* 67/167!; Busoga District: 3.2 km. S. of Nkondo on road to Buyende, Galinyanja Swamp, 9 July 1953, *G.H.S. Wood* 814A!; Masaka District: Lake Nabugabo, Aug. 1935, *Chandler* 1366!
KENYA. Fort Hall District: Thika, north of the Thika R., 16 Aug. 1967, *Faden* 67/629!; Nairobi, Thika Road House, July 1951, *Verdcourt* 536! & Nairobi National Park, forest area, 21 Jan. 1962, *Verdcourt & E.J. Brown* 3249!
TANZANIA. Pangani District: Kipumbwi, Serawani, 10 Aug. 1955, *Tanner* 2047!; Dodoma District: 20.5 km. E. of Itigi station, 11 Apr. 1964, *Greenway & Polhill* 11514!; Mbeya District: Ruaha National Park, Magangwe ranger post, 9 May 1972, *Bjørnstad* 1654!; Songea District: about 6.5 km. W. of Songea, 30 Apr. 1956, *Milne-Redhead & Taylor* 9860!
DISTR. U 2–4; K 4; T 3–8; Central African Republic, Zaire, Rwanda, Burundi, Sudan, Zambia, Zimbabwe and Angola
HAB. Seasonally swampy areas, flooded plains, muddy vleis and damp sand in grassy clearings in forest, bushland and woodland, bare areas in grassland with small ephemerals, Cyperaceae, etc.; 30–1740 m.

NOTE. *R. serpiculoides* has occasionally been confused with *Ammannia baccifera* particularly in the flowering state but the fruits are very different, the former having a capsule splitting into valves with fine striations and the latter bursting irregularly with wall finely reticulate; the calyx-lobes are distinctly more acuminate.

7. **R. gossweileri** *Koehne* in E.J. 42, Beibl. 97: 48 (1908); Exell in J.B. 67, Suppl. Polypet: 186 (1929); A. Fernandes & Diniz in Garcia de Orta 6: 95 (1958); A. Raynal in Adansonia, N.S. 7: 541 (1967); A. Fernandes in C.F.A. 4: 172 (1970) & in F.Z. 4: 317 (1978); C.D.K. Cook in Boissiera 29: 106, fig. 23 F–I (1979); Maquet in Fl. Rwanda 2: 484 (1983). Type: Angola, Malanje, Quizanga, *Gossweiler* 1145 (B, holo.†, BM, K, P, iso.)

Amphibious or terrestrial annual or perennial herb; stems often pink, subsucculent, erect or creeping, 3–20(–30) cm. long, simple or sparsely branched, rooting at nodes. Leaves often tinged reddish, narrowly elliptic to almost round, 2–6 mm. long, 1–4.5 mm. wide, obtuse at the apex, ± narrowed to subcordate at the base, decussate, ± sessile. Flowers reddish or greenish, 4(–5)-merous, solitary in the axils, sessile; bracteoles linear-lanceolate to capillary, 0.5–0.7 mm. long. Calyx-tube campanulate, ± 1 mm. long; lobes red, 4(–5), triangular, ± 1 mm. long, 2 opposite lobes slightly wider at base than other pair or 1 wide, 1 narrow and 2 intermediate; appendages absent (Fernandes says very short or absent for *R. minuta*); interjected folds distinct at anthesis. Petals absent. Stamens (1–)2 (–3); filaments 1 mm. long, not exserted. Ovary red, subglobose, 0.75 mm. long; style 0.25–0.5 mm. long; stigma capitate. Capsule ± globose, 1–1.5 mm. long, 3-valved, scarcely or not exceeding the calyx. Seeds brownish (said to have bright red marks, *Norman* 150), subglobose or obovoid, concavo-convex, 0.3–0.5 mm. long. Fig. 15/28–31, p. 53.

UGANDA. W. Nile District: Maracha, Apr. 1940, *Eggeling* 3880!; Kigezi District: Buhara, 29 Aug. 1952, *Norman* 150!; Mengo District: Kampala, Kabaka's [King's] Lake, 4 Sept. 1935, *Chandler & Hancock* 30!
KENYA. Trans-Nzoia District: Kitale, 18 Sept. 1954, *Bogdan* 4296! & Kitale, Prison Dam, Mar. 1967, *Tweedie* 3426! & Kitale, Kenilworth Estate, 30 Dec. 1953, *Jack* in E.A.H. 10445!

* *Fide* J.P.M. Brenan adnot. *Chandler* 1366 and confirmed.

TANZANIA.   Rungwe District: 'Mwanjali R. Luwangalala'?, 10 June 1912, *Stolz* 1347!; Songea District: Chipoli, 2 June 1956, *Milne-Redhead & Taylor* 10537! & N. of Songea, by R. Luhira, 23 June 1956, *Milne-Redhead & Taylor* 10889!

DISTR.   U 1–4; K 3; T 7, 8; Senegal to Cameroon, Chad, Central African Republic, Zaire, Rwanda, Burundi, Ethiopia, N. Malawi, Zambia and Angola

HAB.   Muddy parts of *Miscanthidium* swamps, *Sphagnum* bogs, swampy outflows from springs, boggy grassland, pool edges, etc., with Cyperaceae; also in old rice fields; 890–1860 m.

SYN.   *R. urundiensis* A. Fernandes & Diniz in Bol. Soc. Bot., sér. 2, 29: 89, t. 4 (1955); Boutique, F.C.B., Lythraceae: 6 (1967); U.K.W.F.: 155 (1974). Type: Burundi, Mosso, Ruyigi, *Michel & Reed* 228 (BR, holo., MO, iso.)
  R. minuta A. Fernandes & Diniz in Bol. Soc. Bot. sér. 2, 31: 151, t. 1 (1957); Fernandez in F.Z. 4: 317 (1978). Type: Zambia, Mufulira, *Eyles* 8343 (SRGH, holo., K, iso.!)

8. **R. filiformis** (*Bellardi*) *Hiern* in F.T.A. 2: 468 (1871); Koehne in E.J. 1: 167 (1881); Hiern, Cat. Afr. Pl. Welw. 1: 372 (1898), pro parte; Koehne in E.P. IV. 216: 37 (1903); D.A. Webb in Fl. Europaea 2: 303 (1968); C.D.K. Cook in Boissiera 29: 121, fig. 26 A–E (1979). Type: Italy, Piemonte, Vercelli, *Suffrens* (TO, holo., G, K, LE, M, probable iso.)*

Amphibious or aquatic annual or perhaps sometimes a perennial 5–40 cm. tall; stems red or brownish, erect, simple or irregularly branched, sometimes spongy and creeping beneath, often densely tufted and leafless beneath. Leaves sometimes red, mainly 1-nerved, decussate; submerged leaves linear to lanceolate, 0.3–2 cm. long, 0.3–0.8(–3) mm. wide; aerial leaves lanceolate to ovate, (2–)5–9 mm. long, 1–7 mm. wide, obtuse to emarginate at the apex, truncate to cordate at the base; petiole absent or very short. Flowers solitary in the axils, sessile or subsessile; bracteoles 2, linear-subulate, 0.2–1 mm. long. Calyx pink; tube cupular, 1–1.5 mm. long, somewhat accrescent; lobes 4, triangular, equal or unequal, 0.5–0.7 mm. long; appendages absent. Petals white or pale pink, 2–4 or sometimes missing, linear to obovate or round, up to 0.75 mm. long, 0.25 mm. wide. Stamens (1–)2(–3–4); filaments 0.5 mm. long, the red-brown anthers included, borne below the stigma. Ovary ± reddish, globose or ellipsoid, ± 0.8 mm. long; style 0.25–0.5 mm. long; stigma capitate. Capsule red-brown, ellipsoid to obovoid or globose, 2mm. long, as long as or ± exceeding the calyx-lobes, 2-valved. Seeds brownish, semi-ellipsoid, 0.3–0.4 mm. long. Fig. 15/32–35, p. 53.

KENYA.   Nairobi District: Golf Range between Wilson Airport and Army Barracks, just outside National Park, 8 Apr. 1978, *M. Gilbert* 5047!; Masai District: Intona, Lolgorien area, May 1979, *Msafiri* 888 (not seen)

TANZANIA.   Singida District: Iramba Plateau, Kiomboi, 30 Apr. 1962, *Polhill & Paulo* 2262!; Njombe District: Upper Ruhudji [Ruhudje] R., Lupembe, Mar. 1931, *Schlieben* 450!; Songea District: Kwamponjore Valley, 20 June 1956, *Milne-Redhead & Taylor* 10848! & N. of Songea, R. Luhira, near Mshangano fishponds, 15 June 1956, *Milne-Redhead & Taylor* 10815!

DISTR.   K 4, 6; T 5, 7, 8; Mali, Nigeria, Cameroon, Central African Republic, Zaire, Ethiopia, Malawi, Zambia, Zimbabwe, Angola, Namibia, South Africa (Transvaal) and Madagascar; also in northern Italy from 18th Century until about 1912 in rice-fields, almost certainly introduced

HAB.   Seasonal marshy pools, pans on laterite, waterlogged soil of rice-fields; 1000–1620 m.

SYN.   *Suffrenia filiformis* Bellardi in Mém. Acad. Sci. Turin Sci. Phys. 1, Pt. IX (1802–3) & XI: 445, t. 1, fig. 1 (1804)
  Rotala filiformis (Bellardi) Hiern forma *typica* Koehne in E.J. 1: 168 (1881) & in E.P. IV. 216: 37 (1903), *nom. invalid.*
  R. heteropetala Koehne in E.J. 22: 49 (1895) & in E.P. IV. 216: 38 (1903) & V.E. 3(2): 645 (1921); A. Fernandes in F.Z. 4: 318 (1978). Type: Ethiopia, Shire, *Quartin Dillon & Petit* 761 (B, holo.†, P, iso.)
  [*R. filiformis* (Bellardi) Hiern forma *hiernii* sensu Hiern, Cat. Afr. Pl. Welw. 1: 372 (1898), pro parte quoad *Welwitsch* 2341 & 2342, *non* Koehne sensu stricto]
  R. debilissima Chiov. in Ann. Bot. Roma 9: 61 (1911). Type: Eritrea, Scimezana Guna Guna, *Pappi* 769 (FT, holo.)
  R. congolensis A. Fernandes & Diniz in Bol. Soc. Brot., sér. 2, 29: 89, t. 3 (1955); Boutique, F.C.B., Lythraceae: 8 (1967). Type: Zaire, Shaba, Lubumbashi, *Schmitz* 1684 (COI, holo., BR, iso.)
  R. heterophylla A. Fernandes & Diniz in B.J.B.B. 27: 106, t. 3 (1957) & in Garcia de Orta 6: 94 (1958); Pohnert & Roessler in Prodr. Fl. SW.-Afr.: 9 (1966); Boutique, F.C.B., Lythraceae: 10 (1967); A. Raynal in Adansonia, N.S. 7: 540 (1967); A. Fernandes in C.F.A. 4: 171, t. 18 (1970) & in F.Z. 4: 318 (1978). Type: Angola, Pungo Andongo, near Lake Quibinda, *Welwitsch* 2342 (LISU, holo., BM, iso.)

* I can find nothing at Kew collected by Suffrens but several sheets from the type locality collected by others.

NOTE. I have followed Cook's broad concept of this species which was based on a survey of almost all the material available. Various combinations of aerial and submerged leaves and different sizes of the latter give rise to some very diverse looking specimens, e.g. the wide-leaved *R. congolensis* represented by *Milne-Redhead and Taylor* 10815 and *Polhill & Paulo* 2262 of the specimens cited above. Fernandes & Diniz restrict the name *R. filiformis* to the Italian populations. I am not entirely certain of the identification of the Nairobi specimen *M. Gilbert* 5047 from wet flushes and ponds on grassy slopes. It was not seen by Cook.

V.C. Gilbert 5267 (**T** 5, Rungwa Game Reserve, Camp 2, 25 Jan. 1969), a sterile specimen, is the "*congolensis*" form of this species.

9. **R. capensis** (*Harv.*) *A. Fernandes & Diniz* in B.J.B.B. 27: 105 (1957) & in Garcia de Orta 6: 94 (1958); Boutique, F.C.B. Lythraceae: 8 (1967); A. Raynal in Adansonia, N.S. 7: 540 (1967); A. Fernandes in C.F.A. 4: 170 (1970) & in F.Z. 4: 317 (1978); C.D.K. Cook in Boissiera 29: 125, fig. 27A–O (1979). Type: South Africa, Drakensberg, *Cooper* 1044 (TCD, holo., K, MEL, Z, iso.!)

Annual amphibious caespitose herb, 0.3–10 cm. tall, longest when submerged; stems red, mostly simple above but creeping submerged parts branched. Leaves usually decussate or occasionally in whorls of 3, submerged linear, aerial narrowly oblong to ovate or upper linear-lanceolate, 0.5–1.5 cm. long, 0.5–1.2 mm. wide, submerged tapering to a slender emarginate apex, aerial obtuse to emarginate at the apex. Flowers minute, 4-merous, solitary in axils, subsessile; bracteoles 2, scarious, minute, 0.1–0.25 mm. long. Calyx-tube campanulate, 0.3–0.5 mm. long, not accrescent but splitting in fruit; lobes 4, triangular, 0.2–0.4 mm. long; appendages absent. Petals absent. Stamens 2; filaments very short, ± 0.1 mm. long, the anthers included within the calyx-tube. Ovary globose, becoming ellipsoid, 0.4–0.7 mm. long; style 0.1–0.3 mm. long; stigma subcapitate. Capsule red, ellipsoid, 1–1.3 mm. long, about twice as long as calyx, opening by 2 valves. Seeds light brown, semi-ellipsoid, concavo-convex, 0.4 mm. long. Fig. 15/36–39, p. 53.

TANZANIA. Dodoma District: Itigi, E. of Bangayega, 30 Dec. 1925, *Peter* 33758 & Lake Chaya, 3 Jan. 1926, *Peter* 34009B & 34012 & 4 Jan. 1926, *Peter* 34143B
DISTR. **T** 5; Cameroon, Zaire (Shaba), Zimbabwe, Angola, E. South Africa and Madagascar
HAB. Presumably lake-sides in dry thicket; 1240–1300 m.

SYN. *Suffrenia capensis* Harv., Thes. Cap. 2: 56, t. 189 (1863)
    Rotala sphagnoides H. Perrier* in Not. Syst., Paris 14: 308 (1953) & in Fl. Madag. 147, Lythracées: 14, fig. 2/4–6 (1954). Types: Madagascar, various specimens from Antsirabe, Andringitra and Firingalava (P, syn.)
    R. robynsiana A. Fernandes & Diniz in B.J.B.B. 27: 110, t. 4 (1957); Boutique, F.C.B. Lythraceae: 10, t. 1 (1967). Type: Zaire, Shaba, Dilolo, *Freyne* 48 (COI, holo., BR, iso.)

10. **R. lucalensis** *A. Fernandes & Diniz* in Bol. Soc. Brot., sér. 2, 31: 153, t. 3 (1957) & in Garcia de Orta 6: 94, t. 3 (1958); A. Fernandes in C.F.A. 4: 170, t. 17 (1970); C.D.K. Cook in Boissiera 29: 130, fig. 28E–H (1979). Type: Angola, Malange, Duque de Bragança Rianzondo, near falls on R. Lucala, *Gossweiler* 11818 (COI, holo.)

Perennial herb to 25 cm. tall; stems caespitose, 4-angled, ascending, lower parts creeping, branched and rooting at the nodes, upper parts ± erect, the lower nodes with bases of fallen leaves somewhat spinulose. Leaves decussate, sessile, the lower (usually submerged) and intermediate narrowly lanceolate, 2.5–3 mm. long, 0.6–1 mm. wide, truncate or ± bifid at the apex (sometimes eroded away *fide* Fernandes), narrowed at the base, upper lanceolate, narrowly elliptic, oblong or ± ovate, 4–8(–10) mm. long, 1–1.5(–4) mm. wide, obtuse at the apex, attenuate or slightly cordate at the base. Flowers solitary and sessile in the axils; bracteoles white, linear, scarious, 0.6–1 mm. long, slightly exceeding the calyx-tube. Calyx-tube white or pink, campanulate, ± 1 mm. long, not accrescent; lobes (3–)4, triangular, 0.5 mm. long; appendages absent. Petals 3–4, lanceolate, very small or up to 0.25 mm. long, lanceolate, persistent. Stamens (2–)4, slightly shorter than the calyx-lobes. Ovary ellipsoid, 0.5–0.8 mm. long; style 0.5–1 mm. long, folded in bud. Capsule ellipsoid, 2 mm. long, 2-valved, at length hidden by bases of leaves adpressed to the stem. Seeds brown to black, ellipsoid, 0.75–1 mm. long. Fig. 16.

UGANDA. W. Nile District: Maracha, Dec. 1937, *Hazel* 397!
KENYA. Uasin Gishu District: Kipkarren, Oct. 1931, *Brodhurst Hill* 565!
DISTR. **U** 1; **K** 3; Zimbabwe and Angola

* Cook treats this as a *nom. invalid.* but I can see no reason for this.

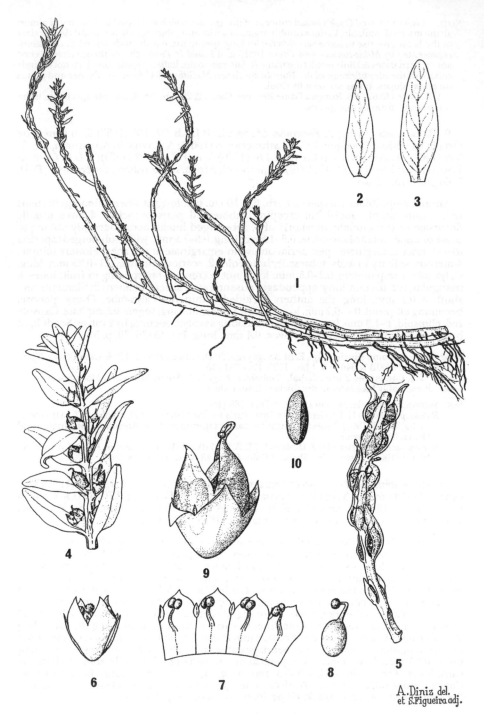

A.Diniz del.
et S.Figueira adj.

FIG. 16. *ROTALA LUCALENSIS*—1, habit, × 1; 2, lower leaf, × 8; 3, upper leaf, × 8; 4, flowering and fruiting stem, × 5; 5, part of stem with leaves adpressed to capsules, × 5; 6, flower with bracteoles, × 20; 7, calyx, spread out, × 20; 8, ovary and style, × 20; 9, calyx with dehiscent capsule, × 20; 10, seed, × 20. All from *Gossweiler* 11818. Drawn by A. Diniz & S. Figueira. Reproduced with permission from Bol. Soc. Brot., sér. 2, 31.

HAB. Seasonally flooded streams; 1290–1770 m.

SYN. *R. wildii* A. Fernandes in Bol. Soc. Brot., sér. 2, 48: 128, t. 15 (1974) & in F.Z. 4: 320, t. 79 (1978).
 Type: Zimbabwe, Mtoka, Makate, Ruins, *Wild* 5662 (SRGH, holo., COI, K, iso.!)

11. **R. stuhlmannii** *Koehne* in P.O.A. C: 285 (1895) & in E.P. IV. 216: 40 (1903); C.D.K. Cook in Boissiera 29: 133 (1979). Type: Tanzania, Biharamulo/Mwanza District, E. Uzinza, *Stuhlmann* 3551 (B, holo.†)

Small gregarious herb with unbranched unwinged stems to 7 cm. long. Leaves decussate, oblong-rhombic, up to 3 mm. long, scarcely 1.5 mm. wide, obtuse at apex, abruptly contracted at the base. Flowers sessile, 4-merous, solitary in the axils along almost the whole length of the stem; bracteoles filiform, almost equalling the calyx-tube. Calyx scarcely 1 mm. long, without appendages; lobes 4, without nerves. Petals absent. Stamens 4, inserted at middle of calyx-tube and eventually equalling the lobes. Style about $\frac{1}{3}$ as long as the ovary. Capsule globose, ellipsoid, smooth, bivalved, ± exserted beyond the calyx-lobes.

TANZANIA. Biharamulo/Mwanza Districts: E. Uzinza, prob. Sept. 1890, *Stuhlmann* 3551
DISTR. **T** 1; not known elsewhere
HAB. Bushland ("pori"); ± 1200 m.
NOTE. Possibly *R. filiformis;* E.G. Baker's very rough sketch of the type (BM!) would fit.

New names validated in this fascicle

Ammannia senegalensis Lam.
 var. **ondongana** (*Koehne*) *Verdc.*, 44
Nesaea burttii *Verdc.*, 18
Nesaea heptamera *Hiern*
 var. **bullockii** *Verdc.*, 30
Nesaea kilimandscharica *Koehne*
 var. **hispidula** (*Rolfe*) *Verdc.*, 32
 var. **ngongensis** *Verdc.*, 32
Nesaea parkeri *Verdc.*, 27
 var. **longifolia** *Verdc.*, 27
Nesaea schinzii *Koehne*
 subsp. **subulata** (*Koehne*) *Verdc.*, 33
Nesaea triflora (*L.f.*) *Kunth*
 subsp. **lupembensis** *Verdc.*, 21

INDEX TO LYTHRACEAE

GEOGRAPHICAL DIVISIONS OF THE FLORA

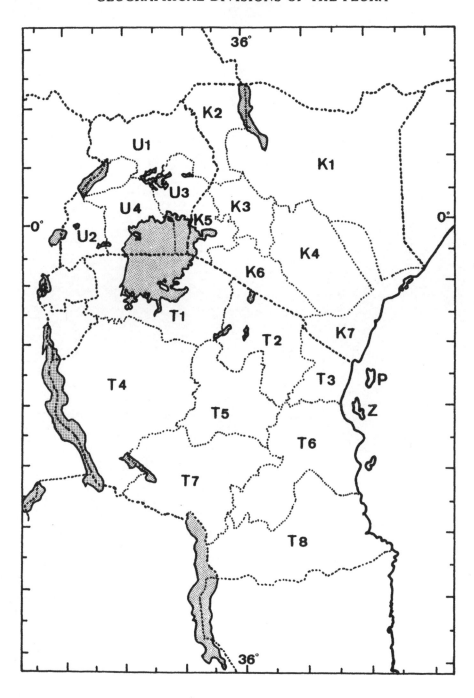

T - #0689 - 101024 - C0 - 244/170/2 - PB - 9789061913665 - Gloss Lamination